W0107476

Research Reports Esprit

Subseries PDT (Product Data Technology)

Project 6457 · InterRob

Edited in co-operation with the European Commission and
the Product Data Technology Advisory Group (PDTAG)

Esprit, the Information Technology R&D Programme, was set up in 1984
as a co-operative research programme involving European IT companies,
IT "user" organisations, large and small, and academic institutions. Managed
by DG III/F of the European Commission, its aim is to contribute to the
development of a competitive industrial base in an area of crucial importance
for the entire European economy. The current phase of the IT-programme
comprises eight domains. Four are concerned with basic or underpinning
and the other four are focused clusters aimed at integrating technologies
into systems. The domains are software technologies, technologies for
components and subsystems, multimedia systems, and long-term research;
the focused clusters cover the open microprocessor systems initiative,
high-performance computing and networking, technologies for business
processes, and integration in manufacturing.
The series *Research Reports Esprit* is helping to disseminate the many
results – products and services, tools and methods, and international
standards – arising from the hundreds of projects, involving thousands of
researchers, that have already been launched.

Springer
Berlin
Heidelberg
New York
Barcelona
Budapest
Hong Kong
London
Milan
Santa Clara
Singapore
Paris
Tokyo

Falk Mikosch (Ed.)

Interoperability
of Standards
for Robotics in CIME

Springer

Volume Editor

Falk Mikosch
Forschungszentrum Karlsruhe GmbH
Postfach 36 40
D-76021 Karlsruhe, Germany

Cataloging-in-publication data applied for

Die Deutsche Bibliothek - CIP-Einheitsaufnahme

Interoperability of standards for robotics in CIME / Falk
Mikosch (ed.). - Berlin ; Heidelberg ; New York ; Barcelona ;
Budapest ; Hong Kong ; London ; Milan ; Santa Clara ;
Singapore ; Paris ; Tokyo : Springer, 1997
  (Research reports ESPRIT : Subseries PDT (product data technology) :
  Project 6457, InterRob ; Vol. 1)
  ISBN-13:978-3-540-61884-3   e-ISBN-13:978-3-642-60609-0
  DOI:10.1007/978-3-642-60609-0

NE: Mikosch, Falk [Hrsg.]; Research reports ESPRIT / Subseries PDT
  (product data technology) / 6457

CR Subject Classification (1991): I.2.9–10, I.5.4, I.6.3, J.6

ISBN-13:978-3-540-61884-3

Publication No. EUR 17233 EN of the European Commission,
Dissemination of Scientific and Technical Knowledge Unit,
Directorate-General Telecommunications, Information Market
and Exploitation of Research, Luxembourg

Typesetting: Camera-ready by the editor
SPIN: 10522444   45/3142-543210 – Printed on acid-free paper

Foreword

The earlier ESPRIT Projects CAD*I (ESPRIT 322) and NIRO (ESPRIT 5109) have made significant contributions to the foundations of Product Data Technology, particularly in the standardisation of product descriptions and robot kinematics in STEP (ISO 10303) and in robotics programming languages. InterRob (ESPRIT 6457), their direct successor, has been building on these results and has extended them to mature applications of Product Data Technology for robotics in high precision manufacturing.

The InterRob approach is based on standardised models for product geometry, kinematics, robotics, dynamics, and control, hence on a coherent neutral information model of the process chain from design to manufacturing. This process thus supports product design, analysis, simulation, robot programming and control by a flexible chain of software modules connected by neutral interfaces. The approach enables the off-line programming of robots relying on CAD product definitions, thus avoiding the much more tedious and inflexible teach-in programming. This capability is a key advantage in one-of-a-kind production.

Applications in plasma spraying of high-precision parts in the aerospace industry and robot welding of complex pipe connections in shipbuilding demonstrate the viability of the approach. The economic success of the methodology is based on the built-in quality control for the information flow and on significant time and cost savings.

InterRob has also made valuable contributions to PDT standardisation, both by relying on existing and evolving standards (ISO STEP AP 203, AP 214) and by extending these models particularly for robotics.

The current volume gives an excellent overview of all significant InterRob results. It is a pleasure to welcome this new book in the Springer series.

Berlin, April 1996
Horst Nowacki
Chairman of PDTAG-AM

Project Partners

BYG Systems Ltd, William Lee Building,
     Highfield Science Park, Nottingham NG7 2RQ, UK

Danmarks Tekniske Universitet, Department of Control and Engineering Design,
     Bygning 424, DK-2800 Lyngby

Forschungszentrum Karlsruhe GmbH (FZK), IAI/PFT,
     Postfach 36 40, D-76021 Karlsruhe

Odense Steel Shipyard Ltd, P.O. Box 176, DK-5100 Odense

Reis GmbH & Co Maschinenfabrik, Postfach 11 01 61, D-63777 Obernburg

Rolls-Royce plc, Manufacturing Technology, P.O. Box 3, Filton,
     Bristol BS12 7QE, UK

SINTEF Informatics, Postboks 124 Blindern, N-0314 Oslo

Contributors

K. Hasund, SINTEF Informatics
P. Hertling, Odense Steel Shipyard
F. Høgberg, SINTEF Informatics
T. Horsch, Reis GmbH & Co Maschinenfabrik
H.C. Larsen, Odense Steel Shipyard
M. Klann, Forschungszentrum Karlsruhe
U. Kroszynski, Danmarks Tekniske Universitet
A. Ludwig, Forschungszentrum Karlsruhe
R. Lutz, Forschungszentrum Karlsruhe
F. Mikosch, Forschungszentrum Karlsruhe
G. Miller, Rolls-Royce
M. Pietrasz, Rolls-Royce
C. Sage, Rolls-Royce
T. Sørensen, Danmarks Tekniske Universitet
P. Sorenti, BYG Systems Ltd
E. Trostmann, Danmarks Tekniske Universitet
S. Whittaker, BYG Systems Ltd

Table of Contents

Annexes

1 Introduction

F. Mikosch
Forschungszentrum Karlsruhe GmbH Technik und Umwelt, PFT
P.O. Box 36 40, D-76021 Karlsruhe, Germany

Information technology and robotics are important instruments to reduce costs, increase product quality and to accelerate the processes within an enterprise, especially product development and preparation of manufacturing. Today serial production e.g. of automobiles is no longer possible without robotics. However, robots are less common in one-of-a-kind and small batch manufacturing, and in manufacturing of variants. This is mainly due to the fact that, facing today's rapidly developing technology, this kind of production requires, much more than serial production, the possibility to replace single components of the production information system by more suitable installations, without impeding the overall information process. Replaceable components like these are, e.g. the CAD system, the programming system, the simulation system, and the robot with its control. These components can, however, only be easily exchanged if they have standardised interfaces. Researchers all over the world work on this subject.

The first main goal of ESPRIT Project No. 6457 InterRob (Interoperability of Standards for Robotics in CIME) described in this book was to close the information chain between product design, simulation, programming, and robot control by developing standardised interfaces and their software implementation for standards STEP (International Standard for the Exchange of Product model data, ISO 10303) and IRL (Industrial Robot Language, DIN 66312) (see Fig.1.1). This is a continuation of the previous ESPRIT projects CAD*I and NIRO, which developed substantial basics of STEP.

The second main goal of the project was to increase the accuracy of off-line programmed robots by improving the conformance of robot simulation with the real world, and by on-line adjustments in robot control to modify actual positions and speed of the robot tool. Effective off-line programming will significantly shorten production preparation, and it will increase product quality as well as productivity and return on investment for expensive installations.

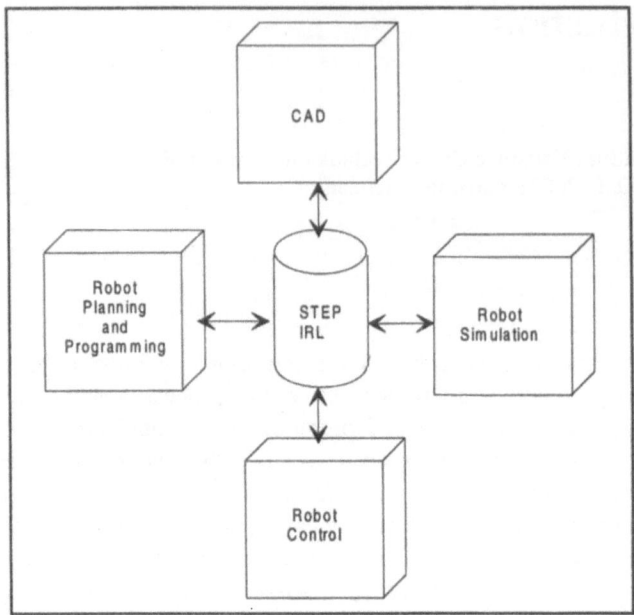

Fig. 1.1. Information Exchange in InterRob

The needs and requirements of one-of-a-kind and small series manufacturers for a consistent bi-directional information flow with open systems in robotics and for the possibility to programme robots fully off-line for precise manufacturing tasks are shown in the pilot applications of the two user companies in the InterRob consortium: Automated off-line programmed plasma spraying and arc welding of parts with a very complex geometry. This new techniques were made possible by combining product data technology, data base technology, and off-line programming and simulation of robots.

The results of the project show an innovative solution for precise manufacturing with fully off-line programmed robots resulting in enormous benefits at the industrial applications. Off-line programming with a much higher accuracy was achieved by the assistance of a simulation tool which takes into account the full geometric information of the work piece and the robot working cell, the kinematic and dynamic behaviour of the robot, calibration data of the individual robot, information about the robot control and the robot tools, and information on process data (welding, spraying etc.).

Fig. 1.2 shows that all data incorporated in the simulation tool are described by STEP or InterRob extensions to STEP, i.e. model schemata which follow the STEP methodology: geometrical data, process data, and robot performance characteristics at any level of available detail.

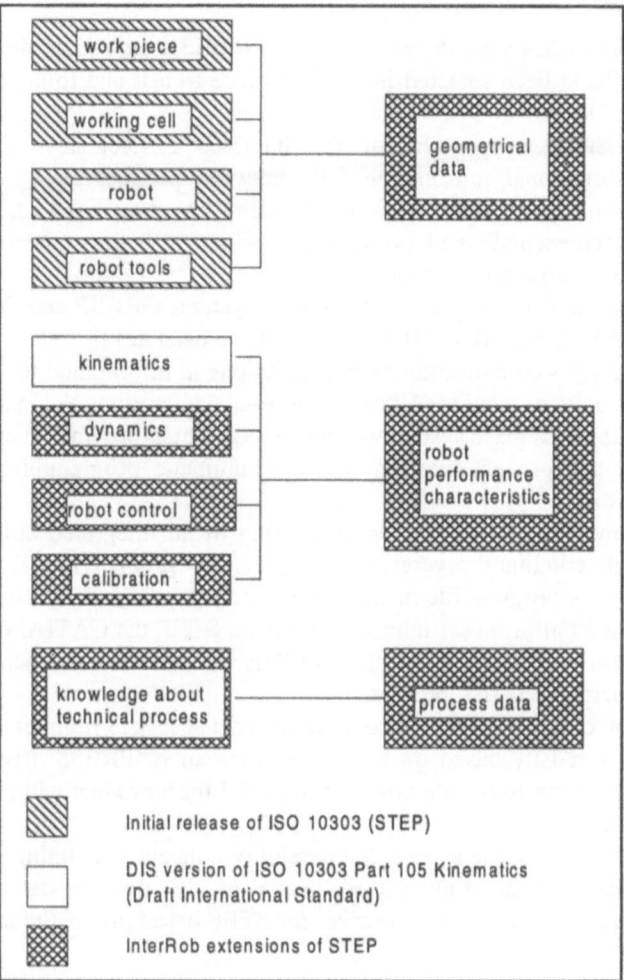

Fig. 1.2. Product Data Transfer in InterRob

The document "Specification of a STEP Based Reference Model for Exchange of Robotics Models" [FZKA-PFT 176/1996] contains a detailed description of all developed schemata. It can be acquired from InterRob Project Management.

Within the framework of InterRob a number of pre- and post-processors for STEP were developed to enable the data exchange between the different systems involved, to prove the conformity of the enlargements (schemata) with the entire STEP standard, and to demonstrate their benefits in industrial applications. For the analysis system ADAMS, the CAD systems BRAVO and CATIA, and for the simulation and off-line programming systems GRASP, KISMET, and ROPSIM a total of six pre-processors and four post-processors with different functionalities according to their tasks were developed. The involved systems and their processor functionalities have been selected in the first place to test and to demonstrate the compatibility of the novel STEP schemata.

The seven European partners of the InterRob Project have developed a consistent bi-directional information flow between product design, simulation, programming, and robot control with open systems based on standards. In addition to the neutral representation of product data with STEP the standard IRL was applied for robot programming and control.

For the simulation and off-line programming systems GRASP and ROPSIM and for robots from Reis and ABB IRL translators have been developed.

The potential for rationalisation by new solutions in information technology can in most cases only be exploited if there is total integration. For handling huge amounts of data, a database system was developed which provides standard access to all product and process information, and automatic programming of special robot applications.

Fig. 1.3 shows the realised system consisting of an integrated object oriented database which combines several subdatabases for product model, equipment, process data, robot program file management, and production data. The system can be operated via a uniform user interface. Through STEP the CATIA, GRASP, and ROPSIM systems are connected. Robots of Reis GmbH & Co and two other robot vendors are operated via IRL programmes.

The InterRob database concept provides uniform access to robotics information. The database is mostly based on STEP and uses an EXPRESS driven database model. All this allows to handle one central model for a product within a company (master model concept).

By this way the database is an interoperability concept integrating standards in the robotics area, facilitating the extension of the data model for specific application areas, and working as a server for STEP based product models within a company.

Important input to the standardisation process of the standards STEP and IRL has been given by the project partners in order to facilitate and improve their use and allowing interoperability.

The InterRob project run within the subprogram "Computer Integrated Manufacturing and Engineering" (CIME) of the ESPRIT Programme (European Strategic Programme for Research and development in Information Technology) supported by the European Commission. It started in January 1993 and lasted for three years.

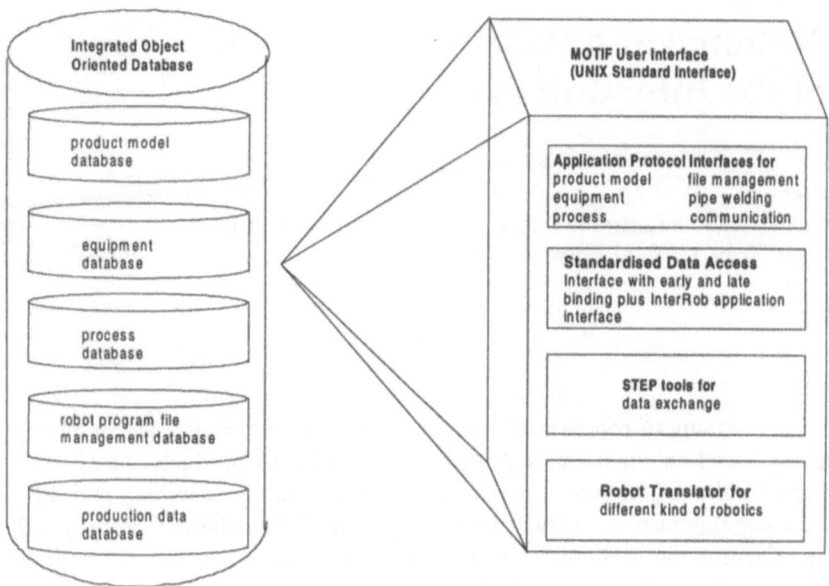

Fig. 1.3. Integrated object oriented database with user interface for uniform access to robotics information at Odense Steel Shipyard

The following companies and research institutes are partners in InterRob:

BYG Systems Ltd, United Kingdom	simulation systems
Danmarks Tekniske Universitet, Denmark	research
Forschungszentrum Karlsruhe GmbH	research and
Technik und Umwelt, Germany	Project Management
Odense Steel Shipyard Ltd, Denmark	user
Reis GmbH & Co, Germany	robot manufacturer
Rolls Royce plc, United Kingdom	user
SINTEF, Norway	assistance to users

In this book the results of the 3-years work (1993-1995) of the above mentioned consortium are reported. It begins in Chap. 2 with a short description of the baseline and rational of the project.

Chapters 3-7 cover the results in detail: Chap. 3 describes the STEP based InterRob interface for the product definition data followed by the description of the implementation of STEP interfaces in InterRob (Chap. 4). Accuracy aspects of off-line programmed robots are dealt with in Chap. 5 under the heading aspects of model fidelity. The two industrial applications and demonstrations of InterRob results are described in Chap. 6. Chap. 7 explains the contribution to STEP and IRL standardisation activities of the project partners. An outlook on the future use and exploitation of the project results concludes this book. Some important details are covered extensively in the Annexes.

2 Rational and Baseline of the InterRob Project

F. Mikosch
Forschungszentrum Karlsruhe GmbH Technik und Umwelt, PFT
P.O. Box 36 40, D-76021 Karlsruhe, Germany

The programming of robots is a bottle-neck in many companies, whose production is mainly based on small batches of frequently changing products. Teach and learn routines for the robots have proven to be difficult, time consuming and the robot and the equipment to be served by the robot are idling during the programming phase. Off-line programming systems with software tools for modelling, task planning, path generation, simulation and programme verification are the obvious solution.

All robot manufacturers use their own programming systems with special programming languages, which differ in various aspects concerning syntax, program structure, and features. Companies using different robots need specialist experts for all systems. An existing robot program for one robot system must be re-programmed in order to be executed by another robot system even though the robots may have a similar kinematic structure.

Robot program portability requires a standardisation of the interfaces between robot planning and programming systems and the robot real-time controller. The project consortium compared the neutral software interfaces

- IRL (Industrial Robot Language)
- ICR (Intermediate Code for Robots)
- IRDATA (Industrial Robot Data)

and chose the IRL standard [DIN 1994] as basis for the development within the InterRob project. None of the three standards has reached so far the status of an accepted European or international standard.

The choice of an appropriate standard interface (neutral file format) for product data description and data transfer from the CAD systems to the robot planning and programming systems was also a central issue in the project formulation. After a long period of development the selected International STandard for the Exchange of Product model data (STEP, ISO 10303) had its initial release during the project time. Baseline for the work done in the InterRob project have been the ESPRIT Projects 322 (CAD*I), 2195 (CADEX), and 2614/5109 (NIRO) which developed substantial basics of STEP. In order to enhance the development of STEP the work was coordinated with the ESPRIT Projects PRODEX and MARITIME and the industry initiative ProSTEP that aims primarily at the implementation of STEP in the European industry.

Precise manufacturing with off-line programmed robots normally requires a considerable amount of on-the-job correction. Simulated and actual robot positions can vary considerably. The main problems occur because robots usually have a good repeatability but a poor accuracy and most simulation systems do not use actual robot control and performance data. The latter information is seen as commercially valuable by the robot vendors and is not generally available to simulation software companies. These problems are being addressed from different points of view.

A project initiated by robot vendors and the automotive industry called Realistic Robot Simulation (RRS) [Bernhardt 1994] is an attempt to provide users with the necessary control characteristics, primarily path planning of a robot in an encoded form via a "black box", thus protecting vendors' proprietary information. A specification for the interface between simulation packages and robots has been published by the RRS consortium and a limited number of robots have had RRS "black box" implementations. RRS, however, does not address robot dynamic performance, which is important for robots moving at high speeds and quickly changing directions.

The InterRob approach aims at the use of open architecture neutral file formats in the STEP methodology. Not only the product structure and geometrical data on work piece, working cell, robot, and robot tools shall be described by STEP files but also all robot performance characteristics like kinematics, dynamics, robot control and calibration data. STEP schemata have been developed accordingly and were positively evaluated together with a schema for process data describing technical processes like plasma spraying and arc welding. This open system concept is user driven and it will take a lot of time and effort to convince robot system vendors in a similar way as the CAD system vendors.

To improve simulation fidelity besides calibrating the robot and the working cell in the InterRob project the ability to bi-directionally communicate data between the robot controller and the simulation package was developed to aid iterative error correction. If detailed dynamic and control data of a robot are lacking it can be attempted to identify dynamic performance characteristics by measuring taught robot movements at different velocities using a laser triangulation system. From the results a correction factor may be derived that can be applied to the off-line created program. In this case corrections will be valid only for specific motions and robot orientations and will not cover the full robot working envelope.

3 The STEP based InterRob Interface for Product Definition Data

U. Kroszynski
Danmarks Tekniske Universitet, Instituttet for Styreteknik
Bygning 424, DK-2800 Lyngby, Denmark

Already a decade ago the ESPRIT project 322, CAD*I (CAD Interfaces) [Schlechtendahl 1989] pioneered joint European efforts for standardisation of neutral data descriptions for exchange of models between CAD systems. The goals in that project focused on geometric shape of mechanical objects in the form of wire-frame models, surface models, and solid models represented in both Constructive Solid Geometry (CSG) and boundary-representations (b-rep). It also addressed the transfer of Finite Element Modelling (FEM) and analysis data, and the use of databases besides textual files as exchange medium.

Among the projects that derived from CAD*I, the subsequent ESPRIT project 2614/5109, NIRO (Neutral Interfaces in Robotics) [Bey 1994], drew upon the results of CAD*I concerning geometry exchange and introduced kinematics as a topical area to be considered for the transfer of models of articulated mechanical assemblies (mechanisms), and, in particular, models of industrial robots.

The information structures developed in these projects, validated by successful model exchange events, were among the first European contributions to the formulation of the emerging STandard for the Exchange of Product model data (STEP) [Owen 1993] that eventually resulted in the ISO-10303 International Standard for Product Data Exchange. In fact, compliance with STEP developments was among the recommendations in CAD*I and a primary strategic goal in NIRO.

Although the kind of articulations considered were only prismatic and revolute (so-called lower pairs), the full topological structure of the kinematics of complete 3D mechanisms was addressed and covered, enabling the successful exchange of a wide variety of industrial robot models between CAD and Robot Programming and Simulation (RPS) systems. The family of models exchanged included mechanisms consisting of simple open kinematic-chains, branched open kinematic-chains, and closed kinematic-chains as found in many modern industrial robots. Moreover, mechanisms could be attached to other mechanisms or be based on the ground. This gave the possibility of modelling articulated tools, like grippers, rotating cassettes, etc. as complete mechanisms mounted on a robotic manipulator.

The geometric representation of the shape of robot arms (link flesh), cell environment (ground), tools, and manipulated objects in NIRO was done by means of faceted-boundary-representations (a general, yet simple approximation to solid shapes) featuring planar contours of points solely. The reasons for this selection were many-fold: the robot simulation and programming systems available in the project only supported this kind of solid geometry modelling, the area of geometry exchange was well established in STEP and did not justify further research within the scope of the project, and the data structures were simple enough to be implemented without special effort, leaving more resources to the kinematics aspects.

In the product description in NIRO, the STEP information model was upgraded with entities that are characteristic to robots among all possible mechanisms. Concepts like Tool-Attachment-Point (TAP frame), mounted or unmounted robotic tool with one or more Tool-Centre-Points (TCP frames), the indication of actuated pairs in closed kinematic-chains, etc. were among these extensions, in a topical area denoted Robotics.

One of the fundamental aims in NIRO was – besides the STEP neutral product description concerning geometric shape, kinematics topology, and some robotics – to attempt a neutral, generic formulation of robot tasks description as an interface between RPS systems (Robot Programming System) and robots. Investigations were made on the Industrial Robot Data (IRDATA) [DIN 1990], Industrial Robot Language (IRL) [DIN 1994], and Intermediate Code for Robots (ICR) [ISO 1991], at the time candidates for proposals to the ISO.

In InterRob, the focus is on applying the now more stabilised ISO-10303 product description interface concerning geometry and kinematics and to expand to other areas of relevance in robot modelling, as well as to examine aspects of inter-operability with other standards, in particular for topical areas of industrial applications where standards do not exist, or are not yet fully developed. Concerning the robot task interface, IRL was selected for investigation and extension, as needed for actual industrial applications.

Concerning the product description interface, ISO-10303 was selected from the beginning. Several events during the project period led towards the elaboration of the InterRob application protocol [Sørensen 1995].

In the same way as InterRob influenced standardisation activities both at national and international levels, the identification of relevant topical areas in the first phase of the project accounted for parallel efforts in ISO, and developments in other projects both within and outside ESPRIT (e.g. PRODEX [Wapler 1993] and ProSTEP [Ludwig 1993]). An important mutual influence took place with the elaboration of parts of ISO-10303 application protocol AP-214 for machine elements in the automotive industry [ISO 1995a]. Among the topical areas considered in AP-214 (the so called Units of Functionality, or UoF) at the onset of the project, were geometry, kinematics, and robotics. The robotics UoF was removed at a later stage, though.

3.1   Identification of Topical Areas for Data Exchange

Most of the efforts in Work Package 1 (WP1) during the first year of the project were devoted to the identification of concepts and data for the topical areas which should be covered by the InterRob information model. This involved expanding the experiences gained during NIRO to be able to describe more realistic robotic models that take into account the enhanced functionality and development trends in CAD and RPS systems.

For robot models to behave realistically it was necessary to address not only the geometric shape and the kinematics of articulated mechanisms but also to include robotics, arm dynamics, and controller models. These new aspects, not formerly considered under the light of neutral descriptions for inter-system exchange had to be carefully defined and limited to a scope that was on the one hand simple enough to be manageable within the project period and resources, while on the other hand complete enough to be relevant and usable in actual industrial applications. Moreover, the needs of the industrial partners of the project consortium for the applications selected for demonstration of results had to be accommodated as well, in the form of requirements from other work packages.

The Technical Annex of the contract between the ESPRIT management and the project consortium already identified the following topical areas:

- GEOMETRY - Needed for the exchange of shape models representing the robots and their environment, as well as the shape of the products being handled by the robots in manufacturing. Early in the project it was determined that the geometry to be exchanged should be in the form of solid models approximated by faceted-boundary-representations, and face-based-surface-models for a more accurate representation of the products. The paths to be followed by the robot tool tip (TCP) along the surfaces, include advanced 3D non-uniform rational B-spline curves (NURBS).

- KINEMATICS - The representation of the topological connection between mechanical assembly components. The components (links) are articulated at so called joints. The movement constraints of one link with respect to another at a joint are determined by a so called pair. In InterRob five kinds of pairs were considered: Prismatic, revolute, cylindrical, screw, and universal.

- ROBOTICS - This topical area includes the characterisation of robots as a class within mechanisms. The inclusion of entities like TAP-frame, Actuator, Robotic-tool, TCP-frame, etc. belong in this area.

- DYNAMICS - This topical area includes, among other things, the representation of mass and inertial properties of links in robotic manipulators that influence the accuracy of their positioning when instructed to follow a prescribed path at high velocity.

- CONTROL - Modelling the robot controller unit which computes the necessary signals to be sent to the drives that generate the movement. The controller determines, e.g., the robot arms' configuration in singular situations where a particular position and orientation of the tip (TCP-frame) of a tool mounted on the robot's TAP can be attained with more than one arm configuration.

In off-line RPS systems using very realistic animated graphic visualisation in advanced computer workstations, the relevance of properly modelling the geometric shape and the kinematics of articulated assemblies is evident. The motivation for including the other topical areas is that models with perfect nominal geometry and kinematics in such systems often produce results that deviate (some times significantly) from the real robot behaviour. Real robots behave according to the laws of physics while computer models are governed by mathematical representations of these laws which often are simplified approximations that neglect physical effects that might be of relevance.

To illustrate the importance of the other topical areas, consider a case where the operator instructs the robot model to follow an L-shaped path like in Fig. 3.1(a). When the mathematical models include only geometry and kinematics while neglecting dynamics, the infinite forces in the sharp corner are not even detected. The realised path in the real world, however, might look as in Fig. 3.1(b), since the velocity can not have discontinuities.

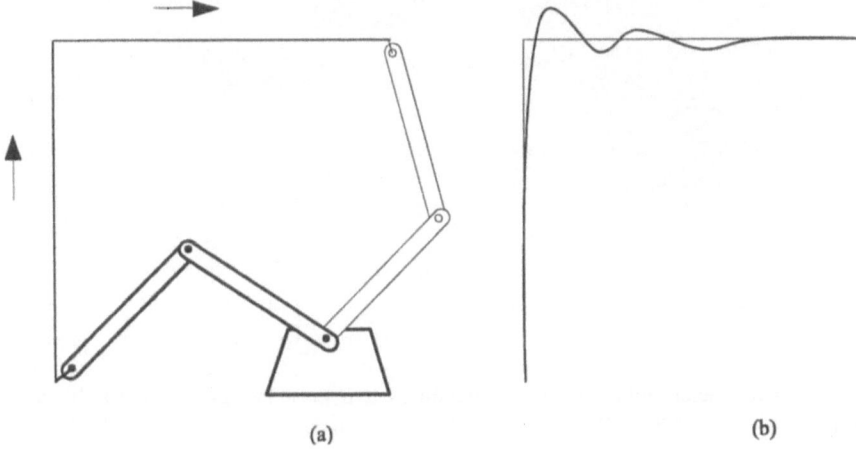

(a)   (b)

Fig. 3.1. (a) Simulated nominal path. (b) Realised path

Although deviations might be small at low velocities, they may well fall outside the allowed tolerances for higher velocities. The whole idea of using RPS systems via models is to be able to detect this kind of deviations beforehand during the simulation rather than at execution time on the robot cell.

The modelling of the robot controller also poses many difficulties. Robot controller architecture and circuitry are well-kept industrial secrets and robot manufacturers are unwilling to disclose specific information. To give an example, one of the classical problems in robotics is that positions and orientations along a programmed path of the robot tool tip (TCP frame) are usually specified in Cartesian space, e.g., as points on, and normals to the surface of the workpiece. A module in the controller (the inverse-kinematic-transformator), computes the actual configuration of the arms (the joint values) in the robot and sends the proper signals to the joint motors to reach that configuration. In special situations as sketched in Fig. 3.2, where the same TCP position and orientation may be attained by two different arm configurations (a) and (b), the controller determines which of the two is realised. The decision is built in the controller circuitry, and depends on the computation method used by the controller manufacturer. For example, it may be based upon an iterative algorithm from the starting configuration and/or upon configuration commands. In the model world this has to be reproduced by either a model of the controller and/or by specifying configuration rules. The model fidelity aspects are investigated in Work Package 2 of InterRob.

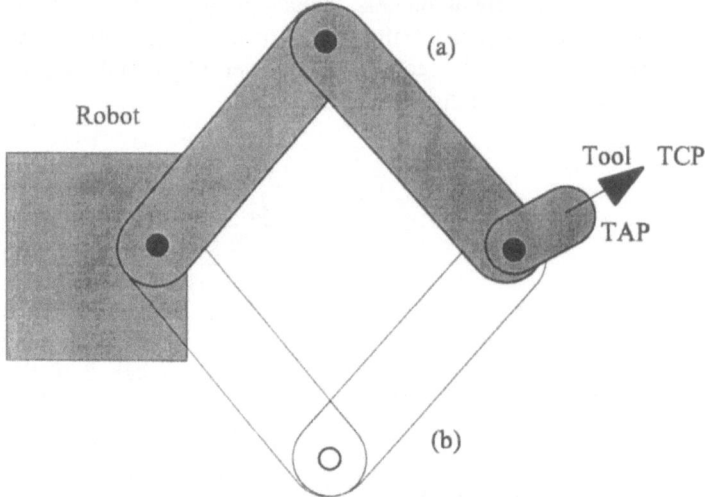

Fig. 3.2. In a special situation, the simulation system must be able to select the same configuration as the real robot

A large amount of the WP1 efforts was aimed at the identification of concepts and data needed for the elaboration of a self contained specification of an information model for robotics that is applicable for the exchange of realistic models for control, programming, and simulation. Although the identification phase was the major activity in WP1 during the first project year, as documented in [Sørensen 1993], it proved to be an ongoing task during the entire project, well beyond the release of the InterRob pilot specification [Sørensen 1994], and even beyond the final InterRob application protocol [Sørensen 1995].

The identification task required a continuous compromise between modelling detail and functional completeness on the one hand and simplicity on the other. The former requirement is an attempt to map the capabilities of existing (and perhaps future) CAD and RPS systems onto the information model. The latter requirement is of paramount importance if relevant results are to be produced within the limited life-span of a project.

The functional completeness of a robotics model needs the inclusion of concepts from many application areas, like shape representation (geometry), kinematics, robotics, dynamics, and control. These were the five major areas stated originally in the Technical Annex. Although the first two, and partly the third were already identified in former projects, they needed a thorough and time consuming review. The last two areas are completely new and were not addressed formerly.

Specific requirements from partners in other work packages, especially from the industrial partners had to be accommodated in the InterRob information schema. Furthermore, compliance with trends in international standardisation imposed the inclusion of general product information, as well as the following requirements:

- Employ ISO-10303 part 11 (EXPRESS [Schenk and Wilson 1994]) as high level data specification language
- Employ ISO-10303 part 21 for mapping to physical textual files as the exchange medium.
- Employ existing constructs from ISO-10303 resources whenever possible (e.g. part 42 for geometry and part 105 for kinematics).

The strategy followed to resolve the conflicting aspects between completeness and simplicity was to consider a minimal set of concepts in the different functionality areas that ensure compatibility with ISO released and draft standards, be present in at least two commercial or R&D systems available in the project (these systems are presented in Chapter 4), and allow for testing the exchange of model features of relevance between these systems.

In the elaboration of ISO-10303 Application Protocol AP-214 functionality areas from many different disciplines had been identified and merged together. At that stage they were identified in a so called Application Reference Model (ARM) as Units of Functionality. Each Unit of Functionality (UoF) deals with a particular aspect of the complete product model and is expressed at the level of information requirements, assertions, and rules to be elaborated upon and detailed to the specification level. Eventually each UoF results in a schema with types, entities, and functions defined in the EXPRESS language. Further restrictions on implementation are the final stage needed before the coding of processor programs for the actual import/export exchange of models between computerised systems can take place.

3.2   The InterRob Application Protocol

The InterRob Application Protocol is a self-contained specification in EXPRESS that merges and integrates constructs from the different sub-schemas with those extracted from ISO-10303 Integrated Resources parts. Eventual rules for the use of these constructs in the context of the application protocol are stated formally while explanations and limitations in implementation are stated informally and illustrated with examples and figures. For convenience, each sub-schema, corresponding to a topical area (or UoF), also presents EXPRESS-G data-structure diagrams, lists of keywords, and alphabetic indices.

Being self-contained, the document [Sørensen 1995] has all the necessary information for implementors of processor programs to code translators to and from the neutral representation without the need to access the original STEP documents.

The InterRob application schema was categorised in sub-schemas, denoted with the same terms used in other STEP related efforts, (e.g. ProSTEP and especially ISO-10303 AP-214) when applicable, and reflecting the different topical areas addressed. The sub-schemas are:

S1 - *General*: A minimal set of product structure information ensuring compatibility with ISO-10303 parts 41, 43, 44, and with related efforts (in particular ESPRIT's PRODEX and the German ProSTEP projects). S1 also includes all constructs which are shared by the other sub-schemas.

Product structure information is an essential part of product data. It provides the constructs that connect the representation data (shape information, kinematics information, etc.) with the data needed to identify the product in the manufacturing environment. Product structure data provide the central post to which all other data related to a product are to be attached. Since the main focus in InterRob is on other product features, a minimal set of product structure data required in any STEP implementation was adopted, as formerly identified in other projects, especially ProSTEP.

Elements like product-structure, representation, representation-context, representation-relationship, measures, units, placement, and transformation are included in this schema. Furthermore, the concepts of representation-map and mapped-item (from ISO-10303 Integrated Resources part 43) are included as a means to structure representation data so it can be used within other representations in the proper context. Its use is restricted to representations that provide shape and positional data, though. This method is used for exchange of symbol-instance occurrences (so called detail-ditto in CATIA, or block-insert in AutoCAD), frequently found in CAD systems.

G3/G4 - *Geometry*: A relatively reduced yet general set of 3D-geometry elements featuring a particular family of surface models (so called face-based-surface-models) and the simplest approximation to solid object shapes by means of polyhedrons (faceted-boundary-representations) respectively. These two categories of geometric shape are relatively simple, yet general enough for the purposes of, e.g., visual realism, collisions detection, etc. and are usually supported by most CAD and some RPS systems. Full compliance with ISO-10303 Integrated Resources part 42 is a dominating aspect in G3-G4.

The G3 sub-schema provides the constructs required for the representation of non-manifold surface models. It contains the geometric and topological information needed for the definition of sculptured surfaces and their boundary curves in the context of non-manifolds. A non-manifold is a surface model that uses topological constructs to define its boundaries and connectivity, and that includes at least two connected-face-sets sharing one face, or two faces sharing one edge.

The G4 sub-schema provides the constructs required for the representation of so called faceted-b-rep models. It contains the geometric and topological information needed for the definition of polyhedral solids by means of point co-ordinates as the only geometric information and point sequences on planar facets as the topological information. The main difference with the more general ISO definition is that in G4 solids may not have voids.

K1 - *Kinematics*: A sub-set of kinematics entities covering complete 3D-mechanisms with open-, branched-, or closed-kinematic-chains. The closed-chains are restricted to planar structures only. The capability of mechanisms attached to other mechanisms is present. The kinematic pairs supported are: prismatic, revolute, cylindrical, screw, and universal. Full compliance with the basic kinematics schema ISO-10303 part 105 (draft international standard) is a dominating aspect in K1.

The basic kinematic structure of articulated assemblies of rigid objects is defined in terms of links, joints and pairs. A link represents a rigid part of the assembly. A joint represents the topological aspect of an articulation between two links, and a pair represents the geometric aspects of the kinematic restrictions of the articulated motion of links at a joint.

K2 - *Robotics*: Basic robotics entities ensuring the exchange of models featuring entities like Workcell, Actuator, TAP-frame, Tool, TCP-frame, etc. and a variety of work-frames describing characteristics of robotic mechanisms and work-cells.

The K2 sub-schema provides the necessary data for describing the planning and programming of operations performed by industrial robots and other programmable mechanisms. It extends the information provided by the K1 sub-schema with aspects which are specific to robotics operations.

A leading principle in this sub-schema was to maximise the inter-operability between processors that are based on the STEP kinematics model (ISO-10303 part 105) and processors that are based on the robotics extensions developed in InterRob. As a consequence, almost all extensions are formulated as additional entities that refer to kinematics entities. Two alternative methods for tool attachment are provided.

K3 - *Dynamics*: A set of constructs describing, among others, mass and inertia properties of robot arms in an approximate fashion.

This sub-schema deals with the dynamics of a system of multiple rigid bodies whose motions are mutually constrained by kinematic restrictions. The behaviour of such a "multi-body-system" under external loads (forces and torques) is governed by Newton's laws.

Computer assisted tools for dynamic analysis simulate Newton's laws on idealised models of multi-body-systems (e.g. by means of concentrated parameters) and produce the time history of the responses (displacements, velocities, accelerations) of the bodies when subjected to input loads. These tools are effectual in providing a more accurate prediction of, e.g., the time needed to perform a robot task, which is one of the major reasons for using RPS systems in the first place.

Two kinds of information are distinguished in the K2 sub-schema: Input information needed to perform a dynamic analysis in the receiving computer system, and output (or analysis result information) to be exchanged together with the input information between systems. Application elements like gear-model, electric-drive, and hydraulic-drive have been foreseen but not specified.

K4 - *Control*: This sub-schema features a partly algorithmic information model associated with controllers. It presents an object-oriented approach for the integrated definition of controller data and functions. The control model covers a description of linear, continuous and discrete dynamic control systems, as well as the general declaration of functions for motion-planning of control reference input. These are compatible with the interface specification of the Realistic Robot Simulation (RRS) project [Bernhardt 1994].

The model-based approach adds a functional description to the "static" model description in EXPRESS by including so-called "services" or member functions to the entities. This way, the entity definition becomes similar to the definition used for objects in object-oriented description languages.

The information model is designed in a generic fashion so it is suited for describing a wide diversity of controllers. Classical controller elements like PID and lead-lag controllers, as well as transfer-function matrices, and linear-state-space-systems are supported.

K5 - *Calibration*: This sub-schema includes a set of constructs intended to introduce some model calibration data in the light of actual comparison between real and simulated robot behaviour. It covers calibration data for placement and for inertial data.

Off-line programming of an industrial robot or any other numerically controlled mechanism relies on the assumption that all relevant data about the mechanism and about its position and orientation relative to the work-pieces are accurate enough for a proper performance of its task. Due to inaccuracies, modelling approximations, and limitations in the RPS systems, this assumption is seldom met.

A method to compensate for these inaccuracies is to calibrate the model by modifying its data in the light of results obtained from specifically designed measurement tests on the real robots. The calibration is successful if it provides more accurate predictions of the motion behaviour of a mechanism at least in some limited part of its entire work-space than its nominal data. This sub-schema provides for the inclusion of calibration data to replace nominal model positioning and mass-properties information.

While nominal data are representative of an entire product class (all robot instances manufactured according to the same design) the calibrated data are associated with a single identifiable instance of such a robot class. Calibration data are also associated with a specific calibration purpose. This purpose is stated in order to achieve a desired accuracy of position and orientation for one or more calibrated frames with respect to their reference frames. The desired accuracy is to be achieved within a certain portion of the entire work-space. This portion is specified either as a region in Cartesian space, or in terms of ranges in the joint-space of the robot.

In the framework of InterRob the investigations contributed in all the above topical areas. The categorisation identifiers S1, G3, G4, and K1 were already partially identified in AP-214 at UoF level, while some EXPRESS constructs concerning geometry were already detailed in ISO 10303-203. The same schema identifiers were also used in, e.g., ProSTEP. The topical areas denoted with identifiers K2, K3, K4, and K5 were first introduced in InterRob and go beyond the scope of AP-214 or any other AP currently under development.

As compared with the original scope envisaged in the Technical Annex at the onset of the project, functionality areas S1 and K5 were identified as important and included in [Sørensen 1993]. The main reason for the inclusion of K5 was that the RPS systems available in the project did not have the capability of mapping manipulator dynamics and control. In order to match simulated model behaviour with real robot behaviour, the idea was to introduce some biasing in the model as a means of calibration.

Since the formal conclusion of the identification phase, three additional schemas have been developed with regard to requirements in WP3 for the industrial applications due as demonstrations in the project. These are the so-called:

A1 - weld data schema,
A2 - spray process schema, and
A3 - weld process schema.

These schemas are not part of the InterRob application protocol [Sørensen 1995] but appear in separate work-documents.

Throughout the development of InterRob, the principle that constructs in a sub-schema should only refer to other constructs in the same or in former sub-schemas was carefully observed. Thus K2 (robotics) constructs only refer to other K2 constructs, or to K1...S1 constructs, but not to constructs in K3 or K4. The InterRob specification developed in WP1 had three major milestones:

1. The results from the identification task [Sørensen 1993] finalised in 1993 but enlarged during the last two years of the project.
2. A pilot specification [Sørensen 1994], released in mid 1994, with the corresponding processor programs and model exchange tests.
3. The final application protocol [Sørensen 1995], a self contained specification of the InterRob information model for robotics released in March 1995, with the corresponding processor programs and model exchange tests.

Between the release of the pilot and the final specifications, ISO-10303 parts 11, 21, 41, 42, 43, and 44 (among others), referred to in the specifications became international standards (IS), while part 105 achieved the Draft International Standard (DIS) status.

These developments in international standardisation needed to be reflected in the InterRob AP. The pilot pre-processor programs to convert models from CAD and RPS systems to the neutral product representation and post-processor programs to decode the neutral representation back to models in the native form required by the CAD and RPS systems had to be updated accordingly. The exchange tests are to ensure that modelling can be resumed in the receiving systems upon model recovery. Thus the final specification is fully compliant with ISO-10303, except of course for the K2-K5 and the A1-A3 schemas which are specific to InterRob. These schemas are detailed in Annex 1.

Since no single computer assisted system can possibly include all the functionality areas in the information model, the InterRob Application Protocol is also categorised in diverse conformance classes. Model exchange events between CAD, RPS, and other computerised systems are restricted to a particular sub-set of functionalities from the entire schema as described in Table 3.1.

Finally, rather than presenting the InterRob Application Protocol itself, the reader is referred to Annex 1 and to document [Sørensen 1995]. The InterRob application protocol has also been published in a limited edition by the project management and is available for STEP related organisations at printing cost[1].

Table 3.1: Conformance classes for the InterRob application schema

Conformance class	Included sub-schemas	Processor programs coded for model exchange tests between systems[a]

---

[1]   For information on project results and the InterRob application protocol kindly contact the Project Management: Mrs. Ursula Frey, Dr. Falk Mikosch, FZK Karlsruhe, PFT, Postfach 3640, D-76021, Karlsruhe, Germany, Phone: (+49) 7247 825 285, Fax: (+49) 7247 825 456, E-mail: frey@pft.fzk.de.

CC1	S1, G3	SISL, KISMET, GRASP, BRAVO3/ROBOT
CC2	S1, G3, A1, A2	GRASP, InterRob database application (ObjectStore)
CC3	S1, G4, K1, K2, K3, K5	CATIA, KISMET, GRASP, BRAVO3/ROBOT, ROPSIM, ADAMS
CC4	S1, K4	ROPSIM, MATLAB/SIMULINK

[a] These systems and results from model exchange tests are presented in Chap. 4.

4 Implementation of STEP Interfaces in InterRob

F. Høgberg et al.
SINTEF Informatics
P.O. Box 124 Blindern, N-0314 Oslo, Norway

The purpose of STEP [ISO 1993] is to enable sharing of product model information between different computer applications. For information to be shared between two systems, an agreement on the contents and semantics of data must be made. In STEP this is accomplished by defining information models in EXPRESS, a formal information modelling language. The semantics of an information model is determined partly by the EXPRESS text (schema) and partly through natural language annotations.

The neutral data carrier may be implemented as textual files or via a database. Using dedicated system to system exchange implies writing processor programs that will translate data structures from the internal native representation of the sending system to the one of the receiving system. With a large number of computer aided software systems, writing these processors becomes an extensive task. Part of the STEP standard is a physical file format, which will act as a neutral format for file exchange. There is no longer a need to write processors for all possible exchange paths as it is sufficient for CA application vendors to write processors to and from the file format defined by STEP. The capabilities of the processors developed in InterRob are summarised in Annex 2.

STEP has also defined a Standard Data Access Interface (SDAI), which enables the creation and manipulation of STEP models in a repository, e.g., a database. SDAI is an abstract programming interface with several language bindings, currently FORTRAN, C, and C++. Applications can share data in a common repository through this interface. Some of the advantages of databases over file transfer is increased security, reduced redundancy and centralised access to data.

In InterRob, both approaches of information sharing have been tested. This is described in the subsequent sections.

4.1 Data Exchange Between Commercial Systems using Physical Files

4.1.1 CATIA

CATIA is a CAD system developed by Dassault Systems in France and distributed world-wide by IBM. Besides the basic configuration featuring 2D and 3D geometry modellers, a series of optional modules are available, like Drafting, Advanced Surfaces, Solid geometry, Kinematics, and Finite Elements. Specialized application modules for manufacturing (like Robotics and Numerically controlled tool-paths), as well as for architecture and plant design (like Piping and Tubing, Structural design, Schematics, and Building design), add to the versatility of CATIA as a high-end CAD system.

CATIA has been implemented on a variety of hardware platforms under diverse operating systems. For instance, at the Control Engineering Institute of the Technical University of Denmark (DTU), CATIA version 3.5 runs on an IBM RISC system/6000 series workstation under AIX/UNIX. This version is coded in FORTRAN 77, C, and Assembler. The DTU particular software configuration supports the base modules, drafting, advanced surfaces, solids, and kinematics. The robotics module license was not purchased though.

Concerning interface capabilities, version 3.5 features the CATIA import/export facility to exchange models with other CATIA installations, an optional CATIA to CADAM converter, and an optional IGES converter. Moreover, FORTRAN subroutine libraries are provided (CATGEO/CATMSP) for accessing the CATIA database via user-coded external programs. The latter interface was the one used in InterRob for coding the CATIA pre-processor to STEP.

The processor queries the CATIA database for a given model, and maps into a STEP file the solids in their polyhedral (faceted-brep) internal representation. It also maps the solid instances (dittos), and complete 3D mechanisms (kinematic-sets). Although CATIA also keeps a CSG binary tree representation of solids in the database model file, the pre-processor only looks for the "evaluated" polyhedral representation. This approximate representation is kept in CATIA for computational ease in display and analysis.

The architecture of the InterRob CATIA pre-processor to STEP is essentially identical to the one used in NIRO [Bey 1994]. It was chosen to hold information in dynamically allocated arrays or tables rather than in direct access files. This is a feature not normally allowed in FORTRAN programming, where array sizes have to be declared beforehand, but made possible by the underlying AIX/C/UNIX environment. Thus, tables can be allocated, extended, or freed as convenient. As compared with the processor developed in [Bey 1994] only the back-end module, mapping the data structures to STEP had to be reprogrammed.

The processor is designed as an off-line program and can be invoked on an alpha-numeric terminal. The user interaction occurs outside the CATIA environment, at the beginning. After a presentation screen the user is prompted to enter:

- The name of the output STEP file (extension .stp),
- The number of digits desired in the mantissa of real numbers [2],
- Whether to apply or not a point-sort algorithm to remove spurious points [3],
- Descriptive text lines, author name, address, and organisation for the STEP file header,
- The CATIA user name to determine database access rights, and
- The CATIA member model name.

The program then accesses the CATIA database and responds on the screen:

```
... Please wait ...
1.Model   accessed...
2.Solids finished...
3.Kinematics copy...
4.Files   released...
... Good-bye   ...
```

Defaults are provided for all prompts. However, the user has to be a registered CATIA user and the member model name must exist in the CATIA database.

The processor corresponds to InterRob conformance class 3 (CC3), and produces a STEP file with the minimal needed product structure information (S1 schema), faceted-breps and their instances (G4 schema), and complete 3D mechanisms kinematics (K1 schema). Moreover, the processor was programmed to interpret and map co-ordinate axes-systems on kinematic sets as additional or TAP frames, and modelled objects with axes-systems in non-kinematic sets as robotic tools with TCP frames (K2 schema). These modelling conventions sort of enhance the functionality available in the DTU implementation with some robotics information as would be found in the (unavailable) CATIA robotics module.

The pre-processor thus generates synthetically entities covering most of the S1-schema, supports the entire G4-schema, the entire K1-schema (with the exception of the Initial_State and Universal_Pair related entities), and some of the K2-schema entities.

Since the merging of two kinematic-sets in CATIA results in one mechanism with more joints, the indication that a mechanism (or tool) is mounted on another mechanism is not feasible. However, some minor hand editing on the STEP file with the separate mechanisms (replacing the ground reference in the mechanism to be mounted with a reference to a TAP frame on the other mechanism), achieves the desired result.

---

[2] Since the STEP file is a textual representation, numbers are decimal coded. Selecting say 4 indicates that real numbers, e.g. point co-ordinates, should be written on the STEP file with 4+1 significant digits, as in +1.4444E-02.

[3] In some unions and subtractions of objects involving tangent surfaces, the resulting CATIA polyhedron contains some points, edges, and facets that need not exist in the combined solid.

A STEP post-processor to CATIA was also coded at DTU, though outside the framework of InterRob[4]. This program reads an InterRob CC3 STEP file, recovers the solid geometry and re-creates the sets needed to define the kinematics. The kinematic-sets, however, have to be defined interactively in CATIA after the exchange event, because no CATGEO interface subroutines are available for this purpose (these routines are delivered with the CATIA robotics module). The DTU post-processor was used to recover models from InterRob STEP files, with a two-fold purpose: To assess the correctness of the DTU pre-processor on the one hand, and to validate the consistency and completeness of STEP files generated by InterRob pre-processors on the other. The recovered models were indistinguishable from the original ones (at least in cycle transfer events CATIA-STEP-CATIA) and modelling could be resumed without problems.

4.1.2 ROPSIM

ROPSIM stands for off-line RObot Programming and SIMulation system. This system is intended for analysis and simulation of the dynamic behaviour of mechanical systems, in particular robots, for given loads and control laws.

The system was developed at the Control Engineering Institute of the Technical University of Denmark by two Ph.D. students, S. Trostmann and L.F. Nielsen, in the framework of the requirements for their doctoral degree with a grant of the Danish Technical and Scientific Research Council.

ROPSIM accepts STEP descriptions of mechanisms (geometry and kinematics). Besides, manipulator dynamics data in the form of mass and inertial properties of links, are to be given. Models of the servo-drives are provided, and data in the form of gains, time constants, etc., as well as friction coefficients have to be specified. The mechanism controller can be modelled by means of a procedure in C language linked to the main simulation code. The loads have to be specified too. The influence of gravity is taken into account. A prescribed nominal path of a point in the mechanism can be specified in the form of a control path in Cartesian co-ordinates. A complete robot program written in ICR [ISO 1991] may also be read as input.

The system was developed on a Silicon Graphics platform, and its software is written in C and C++. It serves as a tool for simulating realistically the realized robot paths and for comparing them with the specified ones, using generic, neutral model descriptions. Since ROPSIM was developed during the NIRO project [Bey 1994], the STEP dialect recognized by the system is no longer valid.

In the framework of InterRob a front-end STEP translator from InterRob to the older NIRO dialect had to be coded. Fortunately, this translation is almost a one to

---

[4] The STEP pre-and post-processors for CATIA are available from IPU, Control Engineering Section, Technical University of Denmark (DTU), DK-2800 Lyngby, Denmark, Att. E. Trostmann or T. Sørensen.

one conversion posing no special problems. Similarly, an IRL translator to ICR was also coded.

Among the InterRob activities using ROPSIM, a simulation of the behaviour of a four-bar pendulum, when left to oscillate under gravity from an initially horizontal extended configuration, led to results comparable to the ones in the ADAMS commercial system at FZK.

A simulation of the dynamic behaviour of a REIS RV-6 robot model, when instructed to follow a nominal Schmid curve, proved qualitatively comparable with actual measurements. A quantitative analysis was not appropriate, since the measurements were conducted without relation with the project and after the simulation itself. A slightly different robot was used, under different configuration settings, and in another spatial position with respect to the plane of the Schmid curve. The simulation could unfortunately not be adapted since no more project resources could be allocated to repeat the task. Further details can be found in Annex 3.

4.1.3 GRASP

GRASP for simulation and off-line programming

GRASP is a general simulation tool using 3D graphical modelling for a wide range of applications from 'discrete event simulation' to the more specific applications of robot modelling and programming. The GRASP system has particular application to the simulation and off-line programming of robotic tasks, and enjoys a large and expanding user base. Experience in this field has demonstrated the need for standardised neutral languages to act as an interface to CAD and native robot language formats.

The processor software for data exchange with GRASP will be made available on all hardware platforms and operating systems on which GRASP is currently available:

- Silicon Graphics
- HP 700 series
- IBM RS6000
- SUN Sparcstations
- DECstations
- HP Apollo
- PC's running SCO UNIX

STEP Processors for GRASP

There are two processors forming the interface between GRASP and the neutral file format STEP, controlling the data flow in each direction. The STEP post-processor converts STEP data files into the native GRASP 'source file' format which can then be input to GRASP. The STEP pre-processor allows models to be output directly from a GRASP session, creating a STEP file in the correct format. These processors have been developed by BYG Systems Ltd. in line with the specification for a STEP based reference model developed in the InterRob project [Sørensen 1995].

The STEP post-processor translates a STEP data file conforming to the InterRob specification, creating a new data file ready for input into the GRASP system. The output files from the STEP post-processor are therefore of the GRASP source file format. The status of the processor at the time of writing is that the InterRob schemas for Kinematics, Faceted-Brep and B-Spline geometries are generally supported. The project end will also see support for basic robotics information such as tool and tcp definitions. Major developments have been made to conform to the new InterRob schemas for welding-specific data which have flowed from the requirements of the pipe welding application at Odense Steel Shipyard. These new schema definitions are detailed as project working documents not included to the main InterRob STEP specification: A1) Weld Data Schema, and A3) Weld Process Schema. The STEP post-processor is a stand-alone piece of software, that is it is not an embedded part of GRASP. It is written in the 'C' programming language, the portability of which allows for transfer to PC platforms, as well as a range of workstations. An example translation of a STEP file conforming to the Weld Process and Weld Data Schemas to the corresponding GRASP native 'source' file is given in Annex 5.

While the STEP post-processor described in the previous section is a 'stand-alone' piece of software, the STEP pre-processor is embedded in the GRASP simulation system. The reason for this is that the format of data in the GRASP internal database is such that it is straight-forward to output GRASP models in the format required by STEP. Translating a GRASP source file into STEP would introduce unnecessary mapping problems and involve additional development of external processor software to perform lexical syntactical and semantic analysis of the GRASP source file. The STEP pre-processor has been developed as an integral part of the GRASP software and will be marketed as additional option available for customers with the system. It is therefore available on any graphics workstation or PC supported by the GRASP system. It is not possible to have access to the pre-processor without a copy of the GRASP software.

All GRASP parameterised solids (CSG-constructive solid geometry) are expanded into a GRASP-specific geometric entity called a GENERAL_MODULE and as such have a one-to-one mapping to the FACETED_BREP STEP entity. Mapping of entities from the STEP face-based surface model follows the same restrictions as imposed on the STEP post-processor described earlier, and mainly resulting from the B-spline entity types supported by the GRASP system. Again the scope of the STEP preprocessor will conform largely to that of the postprocessor as described above and below with reference to the library of test models. The STEP preprocessor of GRASP will also generate a STEP file representation of the calibration data as defined in the K5 schema of [Sørensen

1995] and conformant to the robot calibration 'signature' representation developed under Work Package 2 of InterRob.

GRASP and IRL

When considering the interoperability of robot program descriptions it is often best to start with the level at which the descriptions are defined. In the specific case of GRASP and IRL it is found that both use a high level methodology and a structured programming approach. In many ways IRL takes much of its character from Pascal, GRASP on the other hand uses a proprietary program language chiefly aimed at robotics but containing additional elements for general simulation programming. GRASP's programming language, does not lend itself readily to a comparison to any existing languages. It is a text based language which may be easily created or modified by using the GRASP simulation software or indirectly by an operator using any textual editor. Given that both language descriptions are of a high level and have the same basic requirements in defining robot tasks it is found that the two languages share many of the same features which may be interchanged by simple conversion. With respect to the The compatibility of the IRL *worker subset* [DIN 1994] and the GRASP language allows to translate approximately 80% of IRL constructs to GRASP's programming description. About the same figure holds for the translation from the GRASP language (without the general simulation features not applicable to robotics) to IRL (using all available IRL constructs and definitions).

The IRL to GRASP processor software will take the form of a **translator**. That is, it will take IRL program files as input and produce GRASP format programs (called **tracks**) as output. Since the direction of data flow is from IRL to GRASP the software can be regarded as a *post-processor*. The software will exist as a stand-alone processor in that it will not be integrated with any other software. By typing in the executable file name the operator will then enter the input IRL file name and the output GRASP file name. The software will be written in the C language which will help in making the final product more portable across different hardware platforms and operating systems. In addition, some tools such as those for lexical analysis and file management (also written in C) that were developed in part during the NIRO project may be exploited.

4.1.4 KISMET

KISMET (Kinematic Simulation, Monitoring and Off-line-Programming Environment für Telerobotics) is a software tool for effective planning, simulation, programming and monitoring of remote handling equipments and industrial robots and various types of mechanisms.

The KISMET system was initially developed at the Institute for Applied Computer Science (IAI-CA) of the Forschungszentrum Karlsruhe for applications that support remote maintenance in hazardous environments. The actual KISMET version V4.95.3 requires a Silicon Graphics IRIS 4D Workstation running IRIX 5.3.

KISMET allows various display modes for a real-time synthetic generation of any view of handling and manufacturing cells. Via sensor signals from robot controllers it is possible to update the view parameters, depending on the actual state of the robot.

Mechanical structures are simulated with rigid links. The dynamics module offers a real-time simulation and the modelling of control characteristics of the robot system. Another feature is the modelling and real-time simulation of elastomechanic effects like robot link torsion and bending.

4.1.5 BRAVO/ROBOT

The Applicon BRAVO system has been extended by FZK proprietary software, called ROBOT, for handling kinematics structures. ROBOT has an interface to generate files in the native internal format of the KISMET simulation system. BRAVO/ROBOT is linked to InterRob by using this file as input to the central database RobDB (see Sect. 4.1.7).

4.1.6 ADAMS

ADAMS (Automative Dynamic Analysis of Mechanical Systems) is commercially available from Mechanical Dynamics Inc., USA. It is a Multi-body System Analysis software package for simulating the force and motion behaviour of three-dimensional mechanical systems consisting of rigid bodies, whose mutual movement is subject to various constraints, like kinematic joints. The rigid bodies are represented within ADAMS by their mass properties; shape information is not dealt with by ADAMS.

4.1.7 The STEP Processors for KISMET, BRAVO/ROBOT and ADAMS

The RobDB Database

While the three different systems described above were used in InterRob at FZK (KISMET, BRAVO, ADAMS), it was considered convenient to develop common software and a modular software design for all processors involving these systems. With respect to the STEP specification that was still changing during the InterRob project it was necessary to have a STEP front-end and a STEP back-end that could be adapted easily to the varying releases with a minimum of reprogramming. The consequence was the development of a so-called EXPRESS schema driven STEP front-end. The fact that the commercial systems remained unchanged led to a common data base with multiple access from different sides and with only one STEP processor pair (Fig. 4.1). This central database is called RobDB.

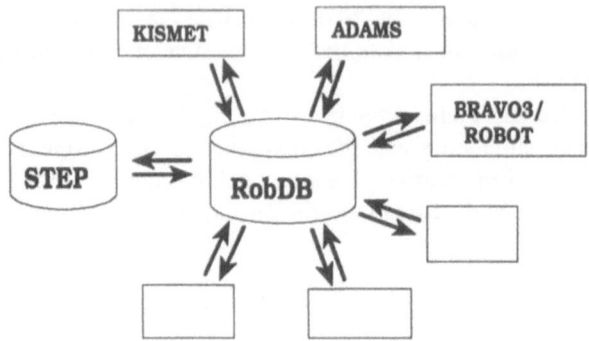

Fig. 4.1. The InterRob RobDB data transfer environment at FZK

The geometric, kinematics, robotics and dynamics data in the RobDB data base are stored as defined in STEP. This means that the contents of the database are a mapping of the information data of the STEP file. The structure of the data - references and organisation - follows the KISMET structure.

As Fig. 4.1 shows, each arrow has the meaning of a processor software module. For each system involved two processor modules are necessary to facilitate the data transfer into and out of each system. To exchange data with another added system, only two more processors are needed.

The data exchange between STEP and RobDB is carried out by the RobDB-STEP pre-processor, which generates a STEP file and the RobDB-STEP post-processor, which reads the STEP data and transforms the STEP entities into RobDB entities. They are based upon the InterRob STEP specification [Sørensen 1995].

The STEP post-processor can be divided into three main modules. The first module is scanning and parsing the STEP file to check the semantic and syntactic correctness. The second module interprets the contents of the STEP entities and creates the RobDB entities and the third module creates the references between RobDB and the STEP file (see Fig. 4.2).

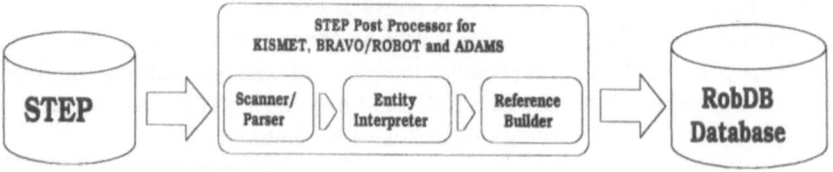

Fig. 4.2. Structure of the STEP post-processor for KISMET, BRAVO and ADAMS

The STEP pre-processor reads the RobDB database and creates a STEP file of its contents.

STEP Pre-processor

The RobDB STEP pre-processor for KISMET, BRAVO and ADAMS is the module which creates a STEP file from the RobDB data model. There exists an EXPRESS schema driven module that allows the creation of single entities on the STEP file as they are defined in the EXPRESS based structure. The references between them are added by the main pre-processor module. A large portion of the InterRob application protocol [Sørensen 1995] was implemented.

STEP Post-processor

Scanner/Parser and Entity Interpreter: For the reason that the Scanner/Parser and the Entity Interpreter of the post-processor are EXPRESS schema driven, it is possible to scan and parse **all** InterRob entities. The entities which should be interpreted must be coded separately within the Entity Interpreter. If, for example, a LINE entity should be created in the RobDB database, this LINE entity must be coded at the automatically created place within the Entity Interpreter module. Consequently, all entities which are not coded, are ignored at the Interpreter. The benefit is, that every InterRob STEP file can be analysed and checked for correctness.

Post-processor Reference Builder: For each entity of the STEP file, that is not referenced by another STEP entity (a so-called TOP entity) the whole hierarchy is built up by the Entity Interpreter. After having created all RobDB entities, the Reference Builder adds the references to these TOP entities. It establishes the hierarchy pointers, assigns information to entities (geometry and frames of a kinematic_link_representation to the corresponding kinematic_links, for instance), resolves or creates relations between two entities, etc. The last task of the Reference Builder is to store all entities in the RobDB database file. It is possible to access this database file by a special access module (read-create-modify-delete access).

4.2   Capability of the STEP Processors for KISMET, BRAVO and ADAMS

The STEP processors for KISMET, BRAVO/ROBOT and ADAMS correspond to InterRob conformance class 3 (CC3) and support the schemas S1 (product structure), K1 (kinematics), K2 (robotics), K3 (dynamics), G3 (NURBS surface geometry), G4 (faceted-Breps) and partly K5 (calibration).

As mentioned above, these schemas are supported by the RobDB STEP processors, while RobDB is the central database within the application.. All three systems use the same kinematic structure defined in the RobDB database.

The differences between the scopes of the three processor back-end pairs depend on the scope of the systems KISMET, BRAVO/ROBOT and ADAMS. KISMET and BRAVO/ROBOT does not use dynamics information within InterRob and ADAMS does not use any geometric shape. While BRAVO/ROBOT uses the same native files as KISMET, the pre-processor supports the same entities. Within InterRob the post-processor back-end for BRAVO/ROBOT supports only geometry.

For a more detailed description of the processors see [Sørensen 1995].

4.3   Data Exchange and Storage by Object Oriented Databases

For one of the InterRob demonstrations, pipe welding at Odense Steel Shipyard (see Sect. 6.2), a computer application has been developed which uses a database for storing and sharing STEP models. The database system used, is the object-oriented database system ObjectStore [ObjectStore 1993].

4.3.1 Object-Oriented Databases

The area of object-oriented databases [Cattell 1991] is a technology which has been under development for more than a decade and is starting to find its use in commercial applications. It is based on research in two areas: persistent extensions to object-oriented programming languages and semantical databases.

In object-oriented programming, the computational model is that of interacting *objects*. An object represents a 'thing' and has a *state* and a set of *services* offered to other objects. These services are called *methods*. During execution of a method, the state of the object may be changed and *messages* may be sent to other objects. The messages will invoke services of the other objects, services which may return a result to the invoking object. The benefits of object-oriented technology are: easier transition from analysis/design to implementation, increased maintainability, and increased reuse of software components. The two best-known object-oriented programming languages are C++ [Stroustrup 1991] and Smalltalk [Goldberg 1983].

Semantical databases (based ,e.g., on Entity-Relationship modelling) allow the explicit definition of entities, sub-entities, relationships, and various constraints. This allows for the development of more complex data models than when using , e.g., the relational model.

CAD systems and simulation programs have traditionally been using files or special purpose databases for storage. With an increased focus on data integration and sharing, general-purpose databases are becoming an interesting alternative. However, traditional databases such as relational databases, have been developed mainly for administrative and financial applications, and do not provide adequate support for CAD systems. Object-oriented databases differ from relational databases in that they are able to store user defined data types (with operations) and build complex data models which may be traversed by advanced methods. In many cases they also eliminate the "impedance mismatch" which is caused by different programming languages for application and database. The architecture and transactional mechanisms of object-oriented databases are also better suited for the typical CAD-user interaction with the database, and will therefore in most cases give better performance.

4.3.2 ObjectStore

ObjectStore [ObjectStore 1993] is a commercial object-oriented database management system [Cattell 1991] from Object Design, Inc. Its object model is based on that of C++ and the programming interface to the database is C++. A Smalltalk [Goldberg 1983] version is also available. An application using ObjectStore, looks very similar to an ordinary C++ program – with some exceptions. Any access to the database has to be within a transaction. Objects are made persistent (i.e. stored in a database) by providing a placement argument (pointer to database) to the *new* operator. Whether objects are persistent or not is thus determined at object creation time. C++ classes which should have the capability of having persistent instances, are marked as such and have to be processed to be stored in a schema database. However, there is no need to maintain separate class hierarchies for transient and persistent classes. Initial access to objects in a database is by database roots – named pointers to objects. Once we get hold of one object, other objects may be reached by traversing pointers. Objects are retrieved from the database by mapping disk pages into virtual memory on demand. If the application program and the ObjectStore server are on different computers, objects are cached on the client side. This improves performance considerably if the working set of the application is smaller than the cache size and contention is low.

4.3.3 STEP Programmer's Toolkit

STEP Programmer's Toolkit [STEP 1992a, STEP 1992b] is a set of EXPRESS tools from STEP Tools, Inc. The part of the toolkit used in the InterRob project, enables import and export of STEP files from/to an ObjectStore database. This requires the use of a compiler which will translate an EXPRESS schema and generate C++ classes for each of the entities in the schema. Another tool will generate an ObjectStore schema description file. The ROSE class library contains functionality for reading a STEP file and instantiating its contents as a C++ object structure (in an ObjectStore database) as well as writing a set of C++ objects (instances of classes derived from EXPRESS entities) to a STEP file. The library interface is not conforming to SDAI, but its functionality covers that of SDAI..

4.3.4 InterRob Application

The InterRob Pipe Program Generation System (IPPG) [Pedersen 1995] consists of a set of ObjectStore databases and applications for supporting the welding of pipes at Odense Steel Shipyard. The databases store information about pipe geometry and welding process as well as equipment data. The different applications allow the user to manage and browse the contents of the databases and also integrates a set of geometric tools.

The product model database stores product model data (for pipes) as well as data related to the manufacturing of the product. Two sub-applications access the product model database. These are the Product Model Management and Welding applications.

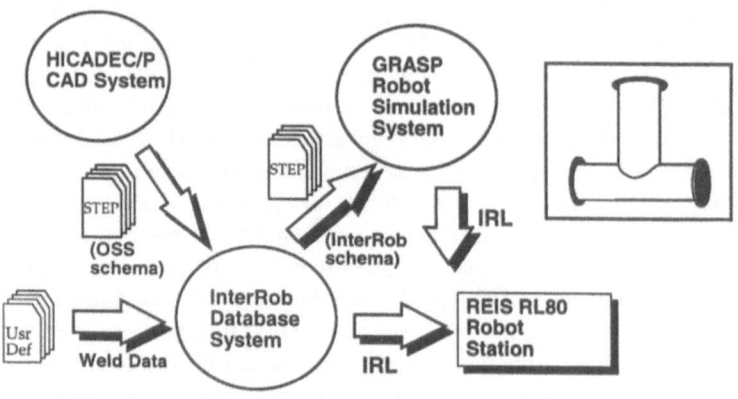

Fig. 4.3. The Inputs and Outputs of the InterRob Pipe Program Generation System

The main usage scenario of the product model database is as follows:

1. Using the Product Model Management application, the user imports a file containing a description of the pipe composite to be welded into the database. A pipe composite will generally consist of a base pipe, a connection pipe (branch), a bend (connected to the end of the base pipe), and flanges. The file is a STEP file based on the schema OSS_piping_model (seeAnnex 5: Example of HICADEC/P STEP file). This schema has been defined to enable data transfer from the CAD system used for drawing pipe arrangements at the

shipyard, HICADEC/P. Behind the scenes, a new EXPRESS model is created for the pipe composite. This EXPRESS model is based on the InterRob schema and contains basic product information and geometry. The wireframe geometry of OSS_piping_model is converted into NURBS geometry. Based on various parameters (e.g. root gap) a groove geometry is generated.

2. The user imports weld data files describing the welding path and weld process parameters for a particular weld. This is done using the Welding application. The weld data file is produced by an application connected to an optical scanner. The scanner is used to scan the groove geometry of pipes and determine the correct positions of welding points. There is one weld data file for each layer in the weld. The weld data file is an InterRob specific and non-STEP conforming text file. When imported into the database, the weld data are mapped to the corresponding entities of the InterRob schema and stored in the database together with the product data.

3. Using the Product Model Management application, the user may now export a file containing a description of the pipe composite, including geometry (NURBS), weld data and weld process data. This file is a STEP file based on the InterRob schema, with entities from sub-schemas S1, G3, A1, and A3 (see Chap. 3 and Annex 5: STEP file from the InterRob database).

The STEP file can then be used by a robot simulation and planning system such as GRASP to generate a robot program for the welding.

4.3.5 The Equipment database

The Equipment database administrates the data concerning information about robots and other equipment in one central database.

Robots need additional facilityies to fulfil their tasks. To show the representation of the management of these facilities within a company (a shipyard for instance), a hierarchical structure can be used (Fig. 4.4).

A 'station' is a specific combination of one or more robots and other equipment (like gantries, tools, etc.), that can do a special task (like welding, flame cutting etc.). The building of different ships (supertanker, container ships, etc.) requires different combinations of stations, the so-called 'plants'. The administration of these plants and stations should be supported by the equipment database.

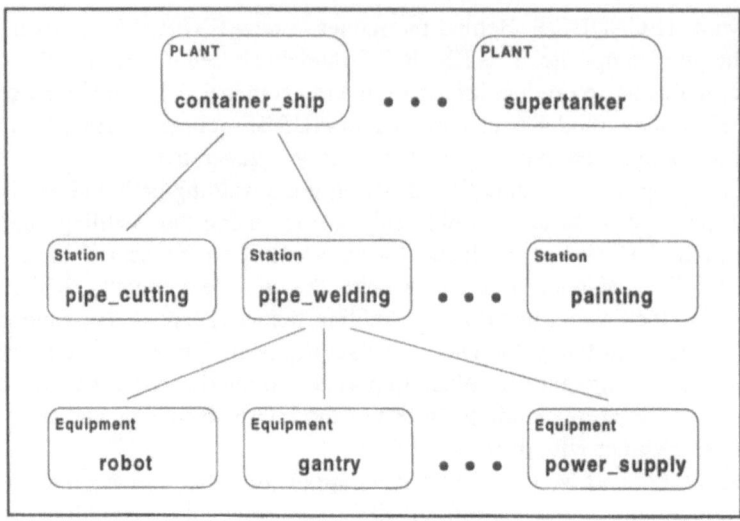

Fig 4.4.   Structure of a ship building facility environment

Architecture of the equipment database

The equipment data can be distinguished by two major aspects (see Fig. 4.5):

1. STEP defined data like geometric, kinematic, and product-definition data, which is defined by the STEP parts 41 - 44, 105 as provided by the InterRob Application Protocol (see Chap.3).

2. additional administrative and technical information, that can not be mapped by existing STEP entities regarding the requirements of the application environment:
   - user data, concerning all user specific information of the facility (location, responsibility, identifier, users manuals, etc.)
   - type data, concerning all type-specific features of the facility (vendor and distributor name, production and sales year, facility manufacturing type)
   - service data, concerning all service aspects of the facility

As the facility administration is an application specific aspect, WP3 tried to remain independent of the common InterRob application schema and it was agreed to implement the equipment schema as an 'add-on' schema. This additional schema is connected with the InterRob schemas to represent the geometric data of the facilities.

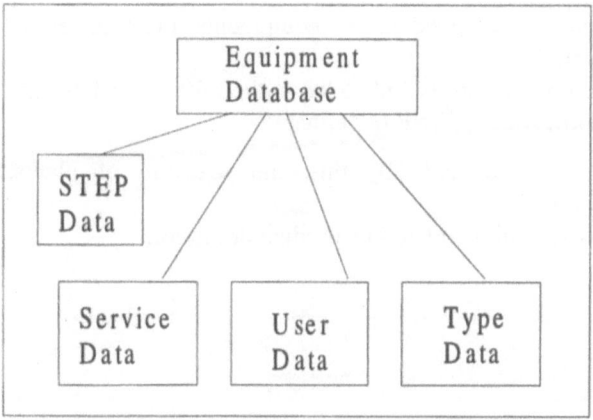

Fig 4.5. Structure of the equipment database

EXPRESS model of the equipment database

One of the base ideas of InterRob is to design STEP based application; that means, that the fundamental data structure is conformant with the proposals of ISO 10303. Therefore the design of the database is realised using EXPRESS with the help of the EXPRESS-G tool 'EXEP' from GIDA, Germany. This tool allows easy generating and testing of EXPRESS code. The resulting EXPRESS file 'equipment_schema.exp' is used to create the data structure for an object-oriented database ('ObjectStore', via the toolkit 'STEPTools'). Fig. 4.6 shows this procedure.

A detailed description of the equipment database and the corresponding EXPRESS model can be found in [Pedersen 1995].

Equipment data schema

The equipment schema is constructed independent from the InterRob Application Protocol [Sørensen 1995]. The design follows as far as possible the rules of ISO 10303. The link to the so-called "STEP data" (product model of equipments ,e.g., robots and their shape) is realised on the level of 'product-definition'. Apart from that the resource schemas 'person_organisation_schema', 'document_schema', and 'date_time_schema' are used. The connection to these schemas is realised by using references (Fig. 4.7).

Import and Export of Equipment data

The equipment database of WP3 application allows the import and the creation of STEP files corresponding to the EXPRESS definition being the base of the database structure. If assigned to the equipments, the transfer of geometry and kinematics is included.

The import/export feature is embedded in the Motif user interface. The user can export to or import from a STEP file either

- the whole database (and, by this, the structure of plants, stations and equipments), or
- only one single equipment and its product definition.

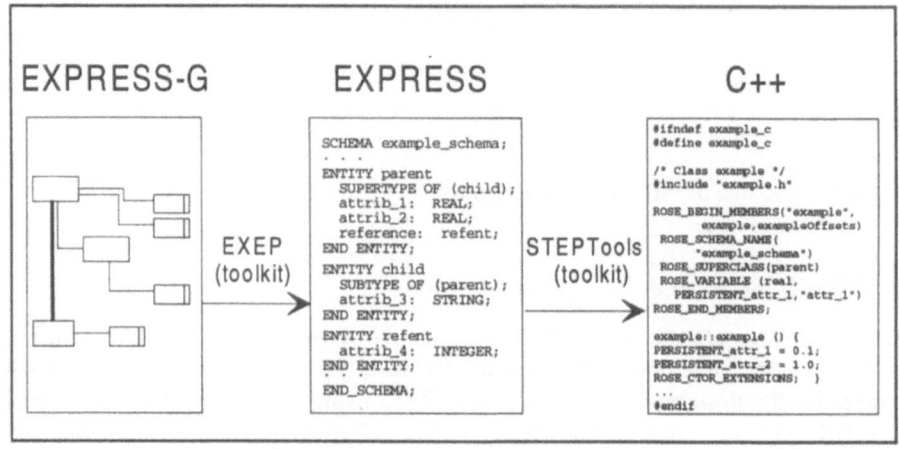

Fig 4.6   Creation of C++ classes from an EXPRESS model.

4.3.6 Interfaces for Object-Oriented databases

Whereas relational databases have a solid mathematical foundation and a (fairly) standardized Data Definition and Query Language (SQL), object-oriented databases have been quite dissimilar both with respect to their underlying model and in functionality and programming interface. In 1993, most vendors of object-oriented databases joined the Object Database Management Group in an effort to define a standard for object-oriented databases. This should make applications source code portable across different object-oriented database systems. Implementations of the standard have been promised by the vendors, but are only starting to arrive (1995).

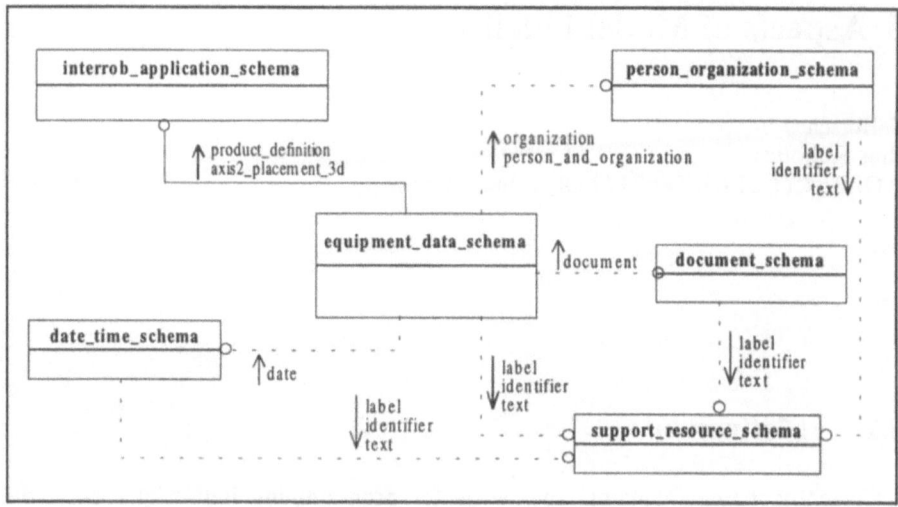

Fig. 4.7. The interaction of the different EXPRESS schemas used by the equipment schema

ODMG 93 may be seen as a competitor to SDAI. There are some important differences between the two standards, though. Whereas SDAI provides the ability to evaluate EXPRESS models for truth value of various logical constraints (local and global rules), the concept is not available in ODMG 93. On the other hand, EXPRESS (and SDAI) does not allow the definition or invocation of methods of entities. SDAI has also been criticized for its lack of a query language.

The ability to model behaviour (e.g. methods) in EXPRESS is also crucial if it is to be used in the context of integration architectures such as Object Management Group's Object Management Architecture. In an environment like this, it will not only be possible to share data, but also services. Work has already started in OMG Special Interest Groups to define object services within different application domains, also in domains covered by STEP. The requirement for behaviour in EXPRESS models have been recognised by STEP and the feature will probably be included in a future version of the EXPRESS language.

5 Aspects of Model Fidelity

T. Horsch
Reis Robotics
P.O. Box 11 01 61, D-63777 Obernburg, Germany

5.1 Motivation

Currently there are mainly two ways for programming industrial robots: the traditional teach-in and off-line programming by using CAR (Computer Aided Robotics) tools like GRASP, ROBCAD or IGRIP. With teach-in programming the programmer drives the robot to the desired positions via a teach pendant and stores them in the application program on the controller. This means, that during teach-in the robot can not be used in the production process. For low batch size production this is an unfeasable approach, since the time spent for programming in relation to production is too high. CAR (Computer Aided Robotics) tools allow to generate programs while the robot is producing using a model representing a class of robots with the same kinematic structure assuming orthogonal joints and predefined link lengths. Nevertheless each robot in its class has its own characteristics due to the manufacturing and assembly process. So the same application program will result in (slightly) different paths in terms of absolute positioning when executed on different robots. Therefore advanced calibration procedures are necessary to match the ideal world to the real world. In the InterRob project advanced methods for error compensation have been developed for both the CAR tool GRASP and the robot control software **ROBOT** Star IV (RS IV).

Off-line programming systems in general allow to input CAD data. Since in current robot controllers only linear and circular interpolators are available, these data need to be converted into a sequence of linear and circular segments. There is no optimal solution for this conversion. In addition this sequence does not guarantee a continuity of the velocity along the path, even if the original data does. In the InterRob project this technological frontier was crossed by developing a new interpolation method based directly on NURBS. Since NURBS is a generic description for any path the existing interpolators can be replaced by the developed one.

The next section describes an error compensation method on the robot controller which allows to execute off-line generated programs without using calibration functionality of the off-line programming system. The following section presents the theoretical background of the new NURBS interpolator and its practical implications. Since this chapter reflects only a part of the InterRob work in Work Package 2, Annex 3 presents further results achieved in the area of simulation of robot dynamics and control.

5.2   Error compensation of robots

This section describes an error compensation method in the cartesian space, which uses a table of measured position errors instead of an analytic error model. This cartesian coordinate correction function can compensate the very wide class of all those errors that may be described as a function of the robot position. The function may be arbitrarily non-linear with the weak restriction of limited variation, i.e. the number of minima and maxima of the error function should not be too high, because otherwise the effort required to determine the data for the error table will become excessive, making the method impractical. The measurements have been taken with the RODYM 6D system from KRYPTON which uses camera-based triangulation as the measurement principle.

The working principle of the cartesian coordinate correction function is based on the fact that the error of the robot end point is for the major part of a systematic nature. There is only a small stochastic contribution to the total error which is due to the limited repeatability of the robot. Measurements on different types of robots have shown that their repeatability is typically between 10 and 100 times better than their absolute accuracy, or in other words the stochastic part accounts only for 1% - 10% of the error budget.

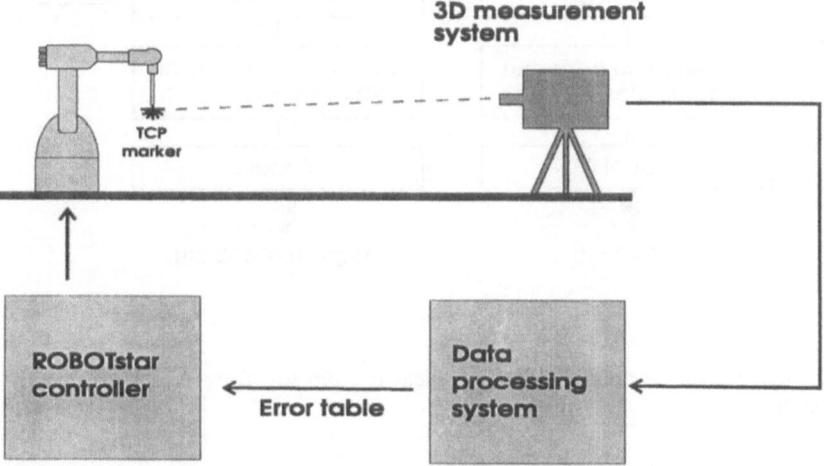

Fig. 5.1. Calibration of the cartesian coordinate correction function with an optical measurement system

Theoretically, it is possible with suitable calibration methods to compensate the systematic part of the positioning error *completely* and *in the whole workspace* of a robot. The cartesian coordinate correction function that has been designed for the ROBOTstarIV controller comes very close to this ideal case. The calibration of this function is based on direct measurement of the cartesian position error in a regular, 3-dimensional grid of so-called "calibration points". The measured error vectors, which constitute the "error signature" of the robot, are stored in a table in the robot controller (see Fig. 5.1). Error compensation within the calibrated workspace is performed by linear interpolation between the table entries; the interpolated error function is applied as a correction term to the forward and inverse kinematic transformations as illustrated by Fig. 5.2. With this data flow the robot programs are stored in "ideal coordinates" and may be exchanged without loss of accuracy with off-line programming systems and other robots that have been calibrated with the same method. The linearising effect of the ROBOTstarIV cartesian coordinate correction function on a robot's coordinate system is illustrated by Fig. 5.3.

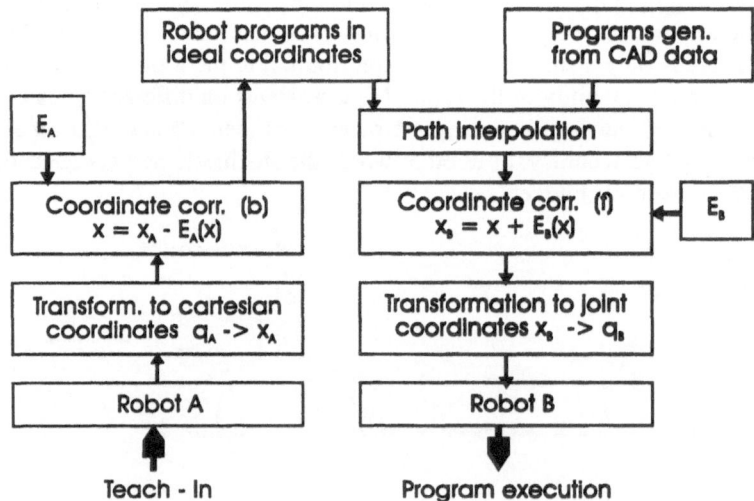

Fig. 5.2. Transformation to and from ideal coordinates using the cartesian correction function.

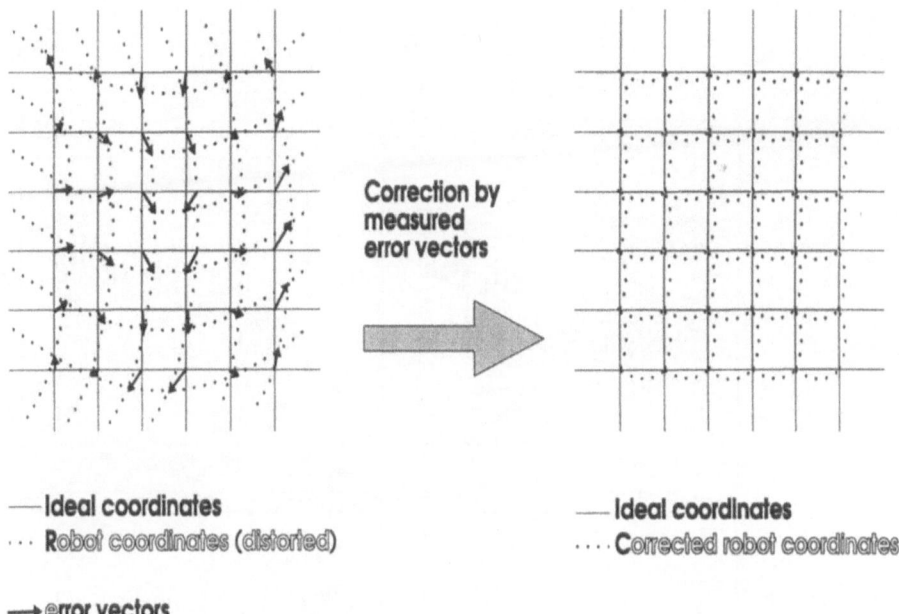

Correction by
measured
error vectors

—— Ideal coordinates
··· Robot coordinates (distorted)

—→ error vectors

—— Ideal coordinates
···· Corrected robot coordinates

Fig. 5.3. Rectification of a robot's coordinate system by the cartesian coordinate correction function.

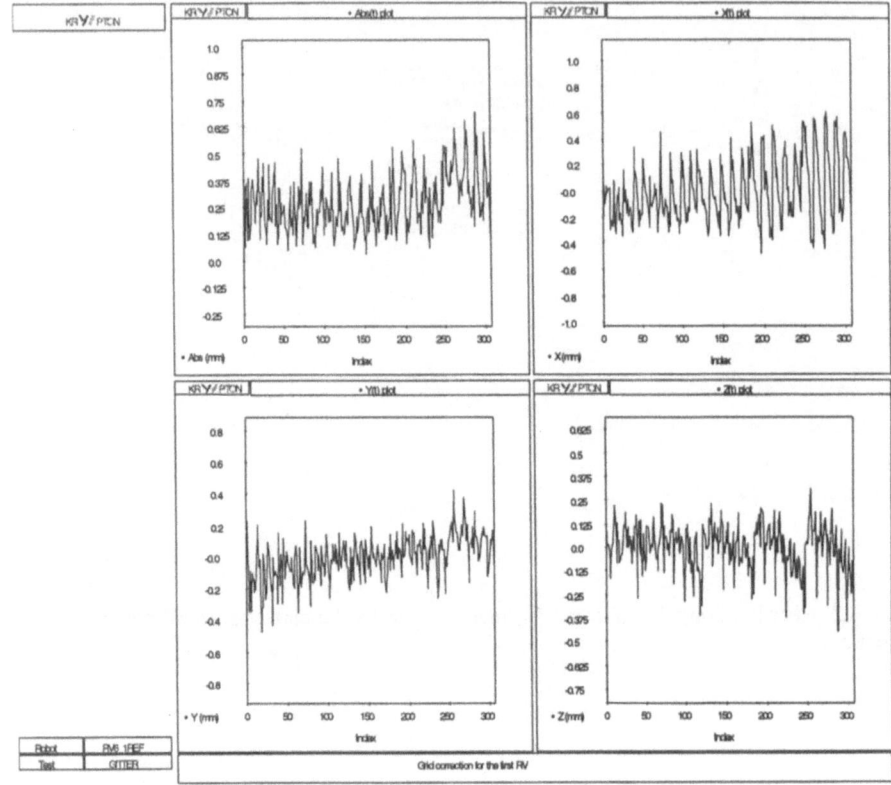

Fig. 5.4. Correction vectors at the calibration points for a Reis robot RV6

Figure 5.4 shows the error vectors at the calibration points for a Reis robot RV6. It represents the absolute value of the errors in mm. This information should be compared to Fig. 5.5 where the calibration points have already been corrected from the controller.

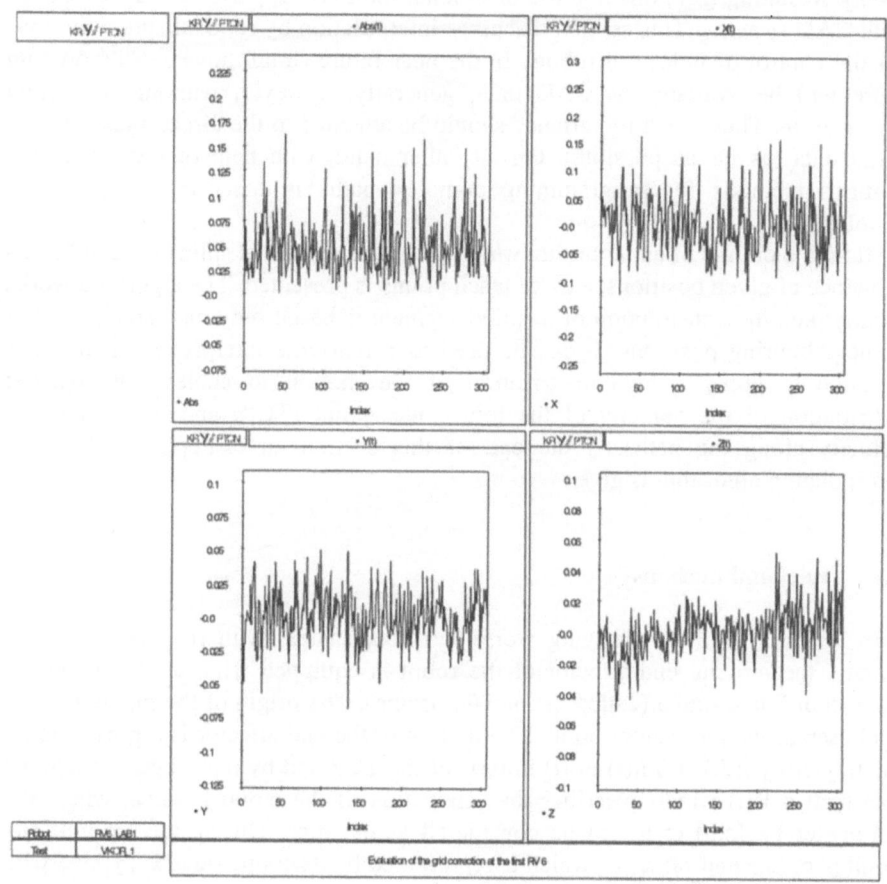

Fig. 5.5. Accuracy of corrected robot at the calibration points

5.3   Spline Interpolation

5.3.1 Introduction

Using spatial rational spline motions it is possible to apply the powerful methods of Computer Aided Geometric Design (e.g., Bézier or B-spline techniques) to problems from kinematics and robotics. Rational motions are defined by the property that the trajectories of the points of the moving object are (piecewise) rational curves, i.e., the trajectories are NURBS curves [Hoschek and Lasser 1993]. Resulting from this, the use of rational motions supports the data exchange with CAD systems. This section discusses interpolation by rational spline motions for the control of industrial robots. In the near future total movements more and more will be available as CAD data, generally, however, without orientation information. Thus, great importance should be attached to the direct integration of CAD data (as far as possible). On the other hand, with help of rational spline motions methods for programming complex paths in *teach-in mode* can be developed.

At first, the outline of an algorithm which generates a rational spline motion from a sequence of given positions, e.g., of teach points is presented. The algorithm works locally, i.e., the construction of a spline segment is based only on a small number of neighbouring positions. It can be used as a real-time interpreter scheme it is possible to apply a reparametrization of the motion to achieve the desired distribution of the velocity of the tool center point (TCP) and of the angular velocity along the path. At the end of this section an example of this new interpolation algorithm is given.

5.3.2 Rational motions

In addition to the underlying world xyz-coordinates (with respect to a *fixed* frame), the moving end-effector of the robot is equipped with another cartesian x'y'z'-coordinate frame (called the *moving* frame). The origin of the moving frame is chosen at the tool center point. The motion of the end effector is represented by the trajectory $m(t) = (m_1(t) \ m_2(t) \ m_3(t))^T$ of the TCP and by the proper orthogonal 3x3 matrix R(t) which describes the orientation of the moving frame, where the parameter $t \in [a,b] \subset R$ can be considered as the time. During the motion, any point $\mathbf{p}'$ on the end-effector (which is represented by its coordinates x_p', y_p', z_p' with respect to the moving frame) runs along the path

$$p(t) = m(t) + R(t) \cdot p' \qquad\qquad (5.1)$$

This equation describes the coordinate transformation $\mathbf{p}' \rightarrow \mathbf{p(t)}$ from the moving coordinate system of the end-effector into world coordinates. If the paths of *all* points $\mathbf{p}$ are (piecewise) rational curves, then the euclidean motion of the end-effector is said to be a (piecewise) rational motion. Piecewise rational (i.e., NURBS) curves are of fundamental importance in Computer Aided Geometric Design where they are usually described in Bézier- or B-spline form, see [Hoschek and Lasser 1993]. Rational motions were discussed in 1890 by G.Darboux at first. We present a simple construction of spatial rational motions.

Let

$$m(t) = \frac{1}{u_0} \begin{pmatrix} u_1 \\ u_2 \\ u_3 \end{pmatrix} \tag{5.2}$$

and R(t) =

$$\frac{1}{D} \begin{pmatrix} d_0^2 + d_1^2 - d_2^2 - d_3^2 & -2d_0 d_3 + 2d_1 d_2 & -2d_0 d_2 + 2d_1 d_3 \\ 2d_0 d_3 + 2d_1 d_2 & d_0^2 - d_1^2 + d_2^2 - d_3^2 & -2d_0 d_1 + 2d_2 d_3 \\ -2d_0 d_2 + 2d_1 d_3 & 2d_0 d_1 + 2d_2 d_3 & d_0^2 - d_1^2 - d_2^2 + d_3^2 \end{pmatrix} \tag{5.3}$$

$$D = d_0^2 + d_1^2 + d_2^2 + d_3^2$$

where $u_0 = u_0(t),..,u_3 = u_3(t)$ and $d_0 = d_0(t),..,d_3 = d_3(t)$ are polynomials in t.

Then, the transformation (5.1) describes a rational motion. It can be shown that all rational motions result from this construction. For a detailed discussion we refer to [Jüttler 1993], where also the relationship between the degrees of the polynomials $u_0,..,u_3,d_0,..,d_3$ and the order of the trajectories is studied thoroughly. The denominator $u_0(t)$ of the trajectory of the TCP is only introduced in order to be compatible with NURBS curves. For the construction of the interpolating spline motion we will set $u_0(t)=1$.

For instance using a non-constant denominator function $u_0(t)$ it is possible to construct exact representations of circular arcs and of other conic sections, see [Hoschek and Lasser 1993]. The *velocity* of the TCP with respect to the parameter t results from

$$\vec{v}(t) = -\frac{u_0'}{u_0^2} \vec{u} + \frac{1}{u_0} \vec{u}'$$

The prime ' denotes the derivative with respect to the parameter t. Note that Eq. (5.3) is the classical representation of a rotation matrix R(t) with help of *Euler parameters*, see e.g. [Bottema and Roth 1979]. As an abbreviation, the four Euler parameters of the rotation matrix R(t) are collected in the four-dimensional *Euler vector*

$$\tilde{d}(t) = (d_0(t), d_1(t), d_2(t), d_3(t))^T$$

The first derivative $d'(t)$ of this vector with respect to t will be called the *Euler velocity* of the motion. The corresponding *angular velocity* of the end-effector is equal to

$$\vec{w}(t) = \frac{2}{\tilde{d}^T \tilde{d}} (\vec{d} \times \vec{d}' - d_0' \vec{d} + d_0 \vec{d}')$$

We do not assume that the Euler vectors $d(t)$ are always normalized (then they would correspond to a curve on the unit sphere in R^4). The use of such normalized Euler parameters would cause an unnecessary doubling of the polynomial degrees required for solving the interpolation problem. The original parameter t of the motion and of the resulting trajectories will not necessarily be the time. After a reparametrization t=t(τ) a motion is realized with the desired distribution of the velocity. The use of rational motions offers the following advantages:

- Rational motions can be handled with help of the powerful methods of Computer Aided Geometric Design (CAGD), for instance using Bézier- and B-spline-techniques. These motions possess a linear control structure which is similar to that of Bézier- or B-spline curves.
- Rational motions generate rational curves and surfaces. The trajectories of the moving points are rational curves (NURBS). Moreover, the surfaces swept out by moving rational curves are rational surfaces [Jüttler 1995], and also the surfaces enveloped by moving planar polygons are rational surfaces.
- Using rational motions enables the direct integration of CAD data to robot controllers.

5.3.3 Construction of spline segment

The representation of the actual spline segment is generated from 4 teach points; the segment starting point and its predecessor as well as the segment end point and its successor. From these points, the tangent at the segment starting and end point will be estimated (see Fig. 5.6). For the first and the last teach point, there will be a special treatment, because there does not exist a predecessor or successor respectively.

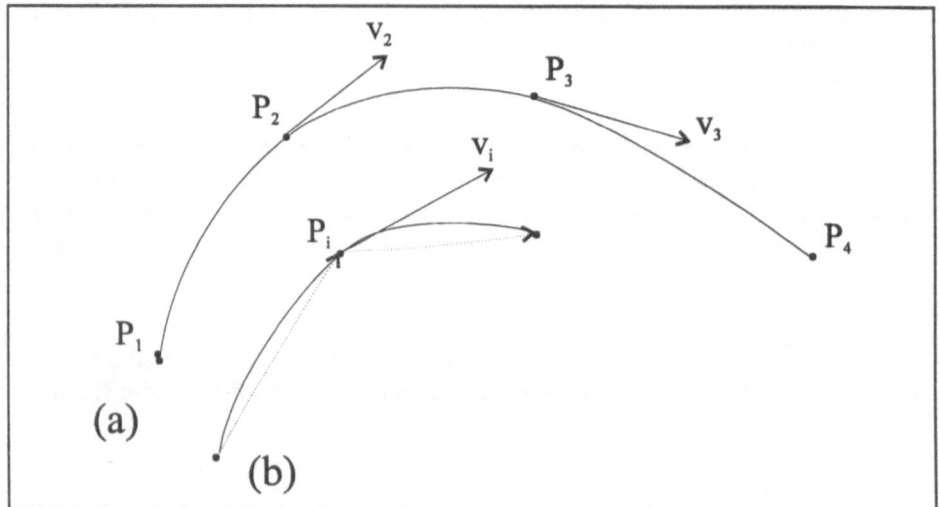

Fig. 5.6. Construction of a path segment

The positions P_1, P_2, P_3 and P_4 are programmed by the operator of the robot. In a first step a centripetal parameterization is performed involving the orientation:

$$t_1 = 0$$

$$t_{i+1} = t_i + \max\{\Delta_{min}, \gamma\|P_{i+1} - P_i\|, 2\delta \arccos(q_i^T q_{i+1})\}$$

$$i = 1,..,3$$

The positive real numbers δ and γ control of the translational and rotational part of the parameter values t_i, whereas Δ_{min} is the minimal allowed difference of adjacent parameter values. q_i describes the orientation in the form of euler parameters for the teach point P_i.

The algorithm then generates from the points $P_1,..,P_4$ and the parameters $t_1,..,t_4$ the tangent v_2 at point P_2 as well as the tangent v_3 at point P_3 (Fig. 5.6 a). The tangent v_i will be *estimated* from the points P_{i-1}, P_i and P_{i+1} (Fig. 5.6 b). In this way the tangent v_2 and v_3 will be generated from the points P_1, P_2, P_3 and P_2, P_3, P_4 respectively, i.e. always the actual point, its predecessor and its successor will be considered for estimation.

From the start and end points of each segment and its tangents control points K_1 - K_4 are calculated, as follows:

$$K_1 := P_2$$
$$K_2 := P_2 + \tfrac{1}{3}(t_3 - t_2) * v_2$$
$$K_3 := P_3 - \tfrac{1}{3}(t_3 - t_2) * v_3$$
$$K_4 := P_3$$
with
$$v_i = \tfrac{1}{2}(\frac{P_i - P_{i-1}}{t_i - t_{i-1}} + \frac{P_{i+1} - P_i}{t_{i+1} - t_i})$$

With the control points K_1 - K_4 and the parameters t_i, a cubic spline for the position is defined, which can be evaluated by means of the Casteljau or de Boor algorithm. In the same manner the orientation is interpolated, which can be hardly visualized (4-D path). Further details of the orientation procedure are omitted here. In the future this interpolator will be used for linear and circular interpolation as well. The construction method of the control points will be similar. Fig 5.7 shows the interpolated positions of the TCP for a robot program with 19 (marked) teach points.

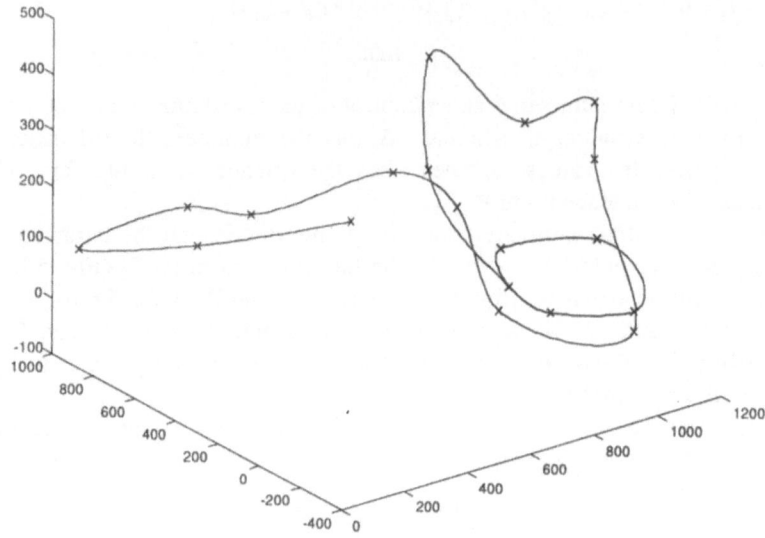

Fig 5.7. Interpolated positions of the TCP for a robot program with 19 (marked) teach points

Acknowledgement

The work which has been undertaken in section 5.3 was supported by B. Jüttler, University of Dundee.

6 Industrial Applications and Demonstrations of InterRob Results

P. Sorenti[1], H.C. Larsen[2], C. Sage[3]
[1]BYG Systems Ltd, William Lee Building, Highfields Science Park
Nottingham NG7 2RQ, United Kingdom
[2]Odense Steel Shipyard Ltd
P.O. Box 176, DK-5100 Odense, Denmark
[3]Rolls-Royce plc,
P.O. Box 3, Filton-Bristol BS12 7QE, United Kingdom

6.1   The Plasma Spraying Application at Rolls-Royce

6.1.1 Introduction

The principal business of Rolls-Royce is the design, development, manufacture, and sale of gas turbine engines and ancillary equipment. It operates through two main groups: The Industrial Power Group, responsible for the design and manufacture of complete power generation, transmission and distribution systems, oil and gas pumping equipment, and marine propulsion units; and the Aerospace Group which specialises in the design and manufacture of gas turbine engines for commercial and military aircraft.

As a world leader in these chosen fields, Rolls-Royce has a reputation for producing high integrity engineering solutions, utilising advanced cost effective manufacturing technology. To maintain its competitive position, Rolls-Royce is continually seeking to improve the performance of its products and to reduce its manufacturing costs.

Manufacturing Technology, a corporate organisation, is part of the Aerospace Group. It has the task of supplying advanced manufacturing technology solutions throughout the company. Based in Bristol, the Robot Application and Systems Group within Manufacturing Technology, is specifically tasked with introducing robotic and automated solutions to a wide range of company products distributed in a number of manufacturing sites. The group has successfully implemented several robot installations, but in a increasingly competitive business, more technically advanced and cost effective solutions are required. Involvement in Esprit project 6457: InterRob, is seen as a very important opportunity to improve our robotic application technology, especially in the field of off line programming.

6.1.2 Overview of Application

The Rolls Royce Business:

The global aerospace market for medium to long range civil aircraft, is dictating that aircraft have a large fare carrying capacity, with low maintenance and operating costs. The new generation of aircraft such as the Boeing 777 is designed to carry up to 370 passengers and cover a distance of 6000 nautical miles using just two propulsion engines. To provide the airlines with the lowest cost of ownership, aircraft are being designed with fewer engines in an effort to reduce maintenance and spares costs. Therefore the engines needed must be large, powerful, yet fuel efficient, and be expected to operate over long periods with minimum maintenance. The Trent engine (see Fig. 6.1 below) selected to power these new aircraft, has the capability to operate in excess of 100,000 pounds of thrust, making it the most powerful engine ever produced by Rolls-Royce. To achieve this level of performance and reliability, many novel design features have been developed.

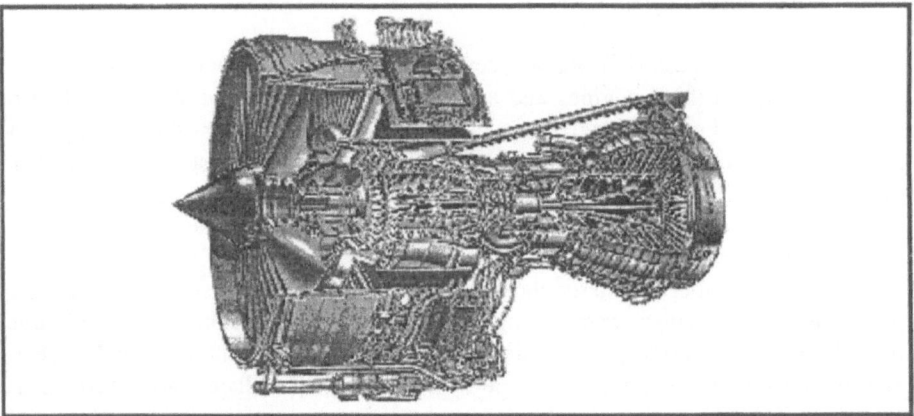

Fig. 6.1. The TRENT Engine

To extract the maximum amount of power from the engine's turbine, and hence produce the maximum thrust, the temperature of the hot combustion gasses entering the turbine must be raised as high as possible. Nozzle Guide Vanes, or NGVs, are designed to direct the hot gas flow from the combustion chamber onto the turbine blades as efficiently as possible. Fig. 6.2 below shows an example component. They are formed into a ring structure, and are located between the combustion chamber and the first stage high pressure turbine blades. During certain operating cycles, the hot gasses striking the gas washed surfaces of the NGV, are significantly hotter than the melting point of the alloy used for their manufacture. Sophisticated internal and external air cooling, plus the addition of a Thermal Barrier Coating (TBC), are used to control the effects of heat.

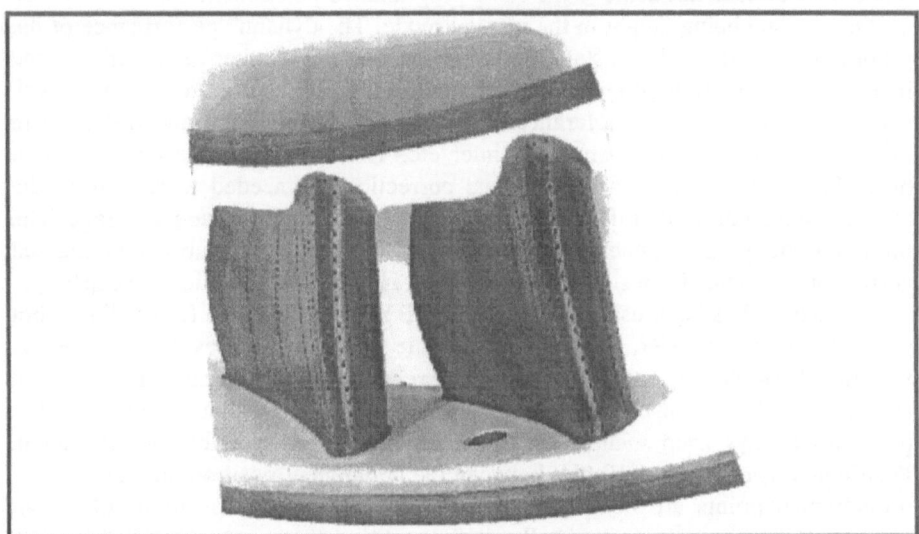

Fig. 6.2. The Nozzle Guide Vane used in the TRENT Engine

Plasma Spraying:

Thermal Barrier Coatings have the effect of reducing the parent metal temperature by up to 170 degrees Celsius. They are formed by plasma spraying a ceramic material onto the previously prepared surfaces of a component. The plasma is formed by supplying a suitable gas mixture around two electrodes contained in a water cooled gun. Using a 50 Kilowatt power source, an electric arc is formed between the two electrodes which is then constricted by the gas mixture to form the plasma. The resulting high velocity plasma jet leaving the gun can reach temperatures in excess of 15,000 degrees Celsius, and velocities of 600 metres per second. The TBC material, usually an Yytria stabilised Zirconia ceramic in the form of a fine powder, is meter fed into the hot plasma stream where it melts and is accelerated towards the component. On impact, the molten particles solidify to form a well adhered protective layer.

Robot Programming:

In the plasma spray cell located in Bristol, an ASEA IRB 2000 robot is used to manipulate the plasma spray gun. Traditionally, robot programs are generated on line at the robot by manual teach and learn methods. To try to determine the contact point of the plasma spray, a pointer representing the stand-off distance of the gun is mounted to the spray gun nozzle. The problem with this approach, is that the robot does not achieve the same programmed positions in automatic mode as it does when being taught in the manual mode. The dynamic performance of the robot is very different to its static performance due to a number of reasons i.e. the inertia effects due to high speed motion, control algorithms used to introduce path smoothing, manufacturing tolerances during robot manufacture, and difficulty in manually teaching points using a pointer etc. The effect of these errors, is that many hours of iterative on line manual correction are needed to reposition the taught points in order that the coating is of the correct thickness and coverage. The initial on line program generation time, plus the time taken for subsequent manual correction, accounts for a significant reduction in the robots productive capacity.

Rolls-Royce has been using GRASP (from BYG Systems Ltd) for off-line robot programming since 1987, but a major obstacle to its more widespread use, is the variation between the component to robot relationship in the simulation model, and the actual relationship in the real world robot cell. Off-line generated robot programs always need manual on line correction to correct for this mismatch. Positional errors of up to 10 mm between the component position and the off line programmed points are not uncommon, though the relative positions of taught points within a program are normally quite accurate.

Rolls-Royce's aim in joining InterRob, was to improve its off line robot programming capability through the acquisition of new technology, and to solve the particular problems associated with the Plasma Spraying application. The target set to prove this new capability was to reduce program generation for a new component by at least 50%, and to produce robot programs off line to a positional accuracy of within 1 mm without the need for on line robot correction.

The technical aspects of InterRob which have contributed to the Rolls-Royce application are:

- Neutral interfaces (STEP) allow CAD defined part geometry to be fed directly into the GRASP 3D Graphical Simulation tool for off-line programming

- The GRASP software has been developed to automatically produce accurate robot programs based on the imported part geometry. GRASP simulates the plasma spray deposition of complex 3D surfaces such that correction can be made off line.

- A neutral robot language (IRL) which allows the program to be transferred to the robot, and to pass calibration and correction data back to GRASP.

- Calibration techniques developed to correct for the robot and cell errors in the simulation model to enhance the accuracy of the off-line programming process.

6.1.3 Description of the Demonstration

The Rolls-Royce demonstration comprised the following features of InterRob technology:

- Calibration: The effects of robot and workpiece calibration are shown by generating a program off line which positions a pointer on a feature of a specially manufactured calibration cube. The robot is then driven to this point both with and without calibration to show the improvement made in positional accuracy.

- Transfer of Product Model Data via STEP: The models of the robot and NGV component are transferred into GRASP using STEP files. Fig. 6.3 shows the data paths used in the Rolls-Royce Industrial Demonstration. The robot model incorporates kinematics and robotics information and the NGV is generated via NURB surfaces.

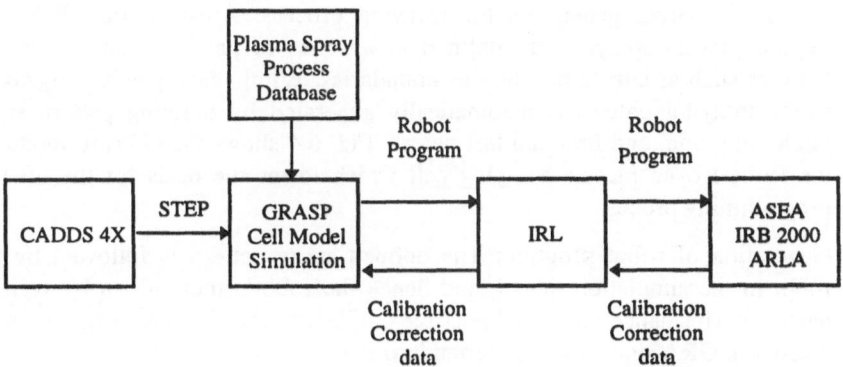

Fig. 6.3. Rolls-Royce Industrial Demonstration Data Paths

- Transfer of robot programs via IRL: Download of off line generated robot programs from GRASP into ASEA IRB 2000 robot, and upload of calibration data into GRASP both using IRL as the neutral interface (see Fig. 6.3).

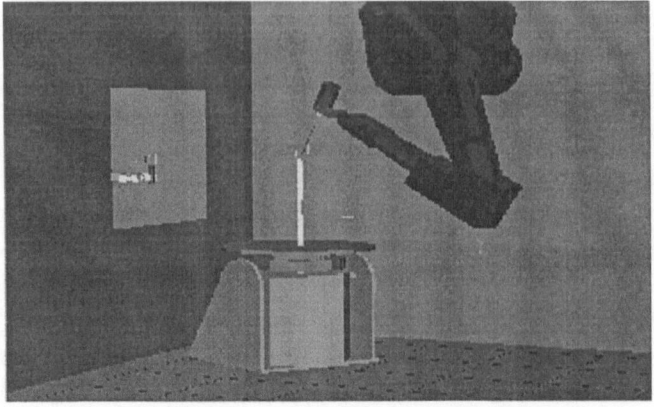

Fig. 6.4. GRASP model of the Rolls Royce Plasma Spraying Cell

- Off line program generation for spraying process: Areas of the NGV that require plasma spraying are defined as a series of patches using geometric features such as curves or edges as boundaries. Within each patch, a zigzag or raster spraying pattern is automatically generated by defining pattern style, pitch, direction, and first and last passes. Fig. 6.4 shows the GRASP model of the Rolls Royce plasma spraying cell which forms the basis for the off-line programming process.

- Generation of robot program: The defined spray pattern is followed by the robot in the simulation model, and hence the robot's motion can be verified against performance limits and possible collisions. The resultant robot program created in GRASP, is then converted into IRL. ·

- Simulation of spraying using a database of spray deposition characteristics: The mathematical model derived from an analysis of earlier plasma spraying tests is used to visualise the deposition thickness by means of a colour or height contour map. Areas indicating thick or thin coatings can readily be seen allowing the programmer to quickly modify the robot path off line. Fig. 6.5 shows an example of the deposition profile generated in GRASP for a pattern used to spray the NGV component.

- Demonstration of plasma spraying: Two patches of the NGV are sprayed using off line generated robot program. Each patch is evaluated to determine deposition position and thickness. From this analysis, a direct correlation to the simulated spraying can be made.

Fig. 6.5. Spraying deposition simulation in GRASP

6.1.4 Results and Experiences

The demonstration of Plasma Spraying at Rolls-Royce has shown that the technology developed in InterRob can be used effectively in an industrial application. Many features from the various Working Groups have been brought together to form a coherent working strategy for future plasma spraying applications.

Calibration of the robot, component, and auxiliary axes has improved the off line programmed positional accuracy from 7 mm to approximately 0,8 mm, when programming to a corner of the calibration cube. This is a significant and valuable result.

The most difficult task in the demonstration, was to obtain sufficiently accurate robot position data that could be used for calibration. To calibrate the robot, component, and auxiliary axes, a suitable pointer is fixed to the robot, and numerous robot positions are recorded about fixed cell features such as the faces of the calibration cube. The robot is therefore used as a measuring tool. The recorded position data, is then fed into the simulation where suitable algorithms calculate the robot Tool Centre Point (TCP), robot signature, and workpiece calibration. To drive the pointer using the robot teach pendant, without damaging the pointer or feature being measured, requires a great deal of manual dexterity on the part of the robot operator. If the position data is slightly wrong, then the calculated TCP, robot signature, and resulting calibration are also wrong.

Transfer of product model and robot model using STEP is very straightforward, however it still remains for Rolls-Royce to obtain STEP output directly from the Company's CADDS (Computervision) system. For this demonstration, the CAD model from CADDS 4X, was converted to STEP using STEFIE (from SINTEF). Robot program transfer via IRL poses no difficulty. The processors developed for InterRob work very efficiently with no detected errors.

One of the main aims of Rolls-Royce, was to reduce part program generation by at least 50% for a new component. For the final demonstration, program generation complete with off-line iteration was accomplished within one week. This compares very favourably with four weeks or more using the previous manual method. This time does not include robot and auxiliary axis calibration time however, since once this is established, it remains unchanged unless the robot is damaged by a collision, new parts are fitted or a major overhaul of the mechanical structure takes place.

The Plasma Spray software, was found to be no more difficult to use than existing GRASP functions. It offers the opportunity to quickly generate various spray methodologies and to visualise the effect of this on coating thickness and coverage. With more use and experience, program generation can be shortened still further, making significant cost benefits for Rolls-Royce.

Following project completion, it is planned to train production personnel in the use of InterRob technology. A new component (a similar NGV part), has already been identified to be the next off line programmed Plasma Sprayed part.

6.1.5 Future application and results

Following completion of Plasma Spraying for a new component, the first new application of InterRob technology will be Vacuum Plasma Spray. This process uses an identical robot to the conventional plasma spray, but it is contained within a vacuum chamber. It is used to spray metallic bondcoats, which form the mechanical interface between the component parent metal and the Thermal Barrier Coating. Other processes being considered are paint spraying, shot peening and NDE (Non Destructive Evaluation). The calibration procedures are already being used on existing robot applications, and the technology developed will be used on a new robot welding process.

6.2   The Pipe Welding Application at Odense Steel Shipyard

6.2.1   Introduction

Odense Steel Shipyard produces large container vessels, product and chemical carriers, supply vessels and fire fighting tugs. Most of the costs and time required for large ships are spent in production, while quality is dependent on the production facilities. In order to raise market shares in an increasingly competitive ship building market, it is company strategy to develop systems that improve efficiency and reduce lead times and costs. In order to obtain this, Odense Steel Shipyard continuously participates in European and nationally funded R&D projects dealing with processes, information technology and related topics. Automation has thus become a keyword, not only to increase productivity but also to give a higher quality and a better base for production planning.

Odense Steel Shipyard already introduced arc-welding robots in production in 1987. The technology was based on experience gained from ESPRIT 595. The welding applications have been continuously developed since then, and the shipyard now controls more than 30 robots in production.

6.2.2   The Application

Until recently only ship assemblies, consisting of plane elements and straight lines have been robot-welded. The future goal is not only to robot weld curved parts, but also highly complex geometries containing advanced multi-layer welds. Pipe-welding is considered to be one of the most complex welding tasks, and as such, was identified as a test application for ESPRIT InterRob.

The pipes are multipass welded with a butt joint in a 3-dimensionally curved geometry with full penetration. The root gap varies from 2 to 5 mm depending on pipe tolerance. All pipes are custom made in various dimensions. This requires both accurate product model data and handling of these from design to production. The programming of the robot must be executed completely off-line and automatically, otherwise programming could take as long as manual welding of a pipe.

Through the InterRob project, Odense Steel Shipyard is developing new solutions for handling product model data, accurate off-line programming and monitoring of production data.

Odense Steel Shipyard processes up to 45000 pipes per year. Each pipe consists, of 2-3 different weld types. A typical pipe is shown in Fig. 6.6 below.

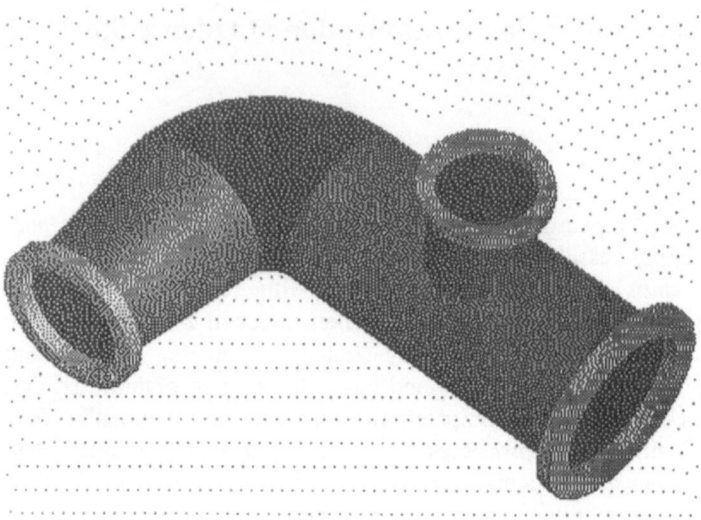

Fig. 6.6. Typical pipe, processed by Odense Steel Shipyard.

On this pipe the three flanges are welded with two layers each, the bend is welded with three layers on each side and the connecting pipe is welded with three layers. For manual welding this gives five hours of welding, but the actual arc time is not more than one and a half hours. The most complex welding is the connecting pipe. Developments within InterRob have concentrated on this welding area.

Pipe production at Odense Steel Shipyard has already partly been automated. Cutting of pipes is done with a numerically controlled flame cutting machine (see Fig. 6.7). This machine provides a cut with an almost equal seam profile around the connection pipe. This seam profile is difficult to obtain using manual flame cutting and it is necessary for the automation of the welding process. As such the automation of the cutting process forms a base for automation of the welding.

Welding of flanges on pipes of smaller dimensions has already been automated, but welding of connection pipes has always been done manually.

The pipe welding application is designed for welding of connection pipes of large sizes at Odense Steel Shipyard. The test application is built in a test laboratory at the shipyard. The pipe welding application consists of a REIS RL80 robot, a Grasp simulation and off-line programming station, a laser scanner system and the InterRob database.

Fig. 6.7. Numerically controlled pipe cutting machine

6.2.3 The Reis 6 Axis Robot with a Turn/Tilt Table

Reis Robotics has developed a new robot for the use in the InterRob project. Prior to this development OSS undertook simulations of the application task in order to evaluate the kinematics structure and to guarantee the reachability of different work pieces. The outcome of this investigation is a gantry system (3 translational degrees of freedom) with an attached vertical arm (3 rotational degrees of freedom). The structure is displayed in the following table including upper and lower limits of each joint.

The demonstration set-up includes a turn/tilt table which is a RDK 1100 table taken from the product range of Reis Robotics. Fig. 6.8 illustrates the on-site installation of the Reis robot cell at Odense Steel Shipyard.

In addition the system is supplied with a coordinate interface to the robot control. This interface enables control of the robot from an external computer on Cartesian level in each interpolation cycle in order to incorporate complex (pre-processed) sensor input data to the system, such as the laser scanner in the OSS demonstration.

Parameter	Axis					
	1	2	3	4	5	6
Joint type	T	T	T	R	R	R
Lower limit				-181 °	-121 °	-359 °
Upper limit				181 °	121 °	359 °
Motion range	2.0 m	1.0 m	1.0 m	362 °	242 °	718 °
Max. speed	1.5 m/s	1.5 m/s	1.2 m/s	270 °/s	270 °/s	270 °/s
Max. acceleration	2.5 m/s^2	3.0 m/s^2	2.4 m/s^2	900 °/s^2	900 °/s^2	900 °/s^2

Table 6.1. Kinematic structure of the REIS RL80 robot.

Fig. 6.8. The REIS RL80 Robot Installation for Pipe Welding at Odense Steel Shipyard

6.2.4 The GRASP Simulation System

BYG Systems Limited develops and sells one of the world's leading 3D Graphical Simulation software tools. One of the company's roles in the InterRob project has been to apply and develop its robot simulation software 'GRASP' for robot modelling and off-line programming in the pipe welding application at Odense Steel Shipyard. BYG has developed software processors to import STEP files for the robot geometry and kinematics (for the REIS RL80 and RDK1100 cell in this instance, see Fig. 6.9 below which illustrates the GRASP model display). These STEP files, conforming to the InterRob STEP schema also contain product model data for the pipe workpiece. This includes the pipe geometry, modelled as NURB surfaces, the weld groove geometry where the pipes are to be welded and also the weld seam geometry (NURB curves) for the three weld passes.

An important advancement for the STEP schema has been the inclusion of the weld process parameters such as weaving frequency, weld speed, arc voltage etc. which can also now be imported from CAD (HICADEC/P to the STEP database and then out to a STEP file) to the GRASP off-line programming system.

The GRASP system supports the modelling, both geometric and kinematic, of robots, their auxiliary equipment and workpieces. It also allows robot programs to be generated, simulated and verified against collisions and performance constraints of the robot. For the InterRob project BYG has developed a high level 'menu' of commands that relate directly to the arc welding process. It allows the user to reference the geometry of the weld curve imported from STEP and then establish the required robot/tool posture along the seam to be welded. The automatic association of welding parameters (weaving data, welding supply power etc.) with points along the weld curve from the STEP file makes the task of creating the robot program off-line very straightforward.

Simple and rapid creation of robot programs directly from CAD data means that the welding task can be quickly evaluated against collisions and other robot performance constraints such as velocity, singularities and joint limits. Once created and verified in GRASP the robot program is then passed to the robot controller via IRL (Industrial Robot Language) using the prototype preprocessor software developed by BYG. The IRL robot program is then passed to the REIS robot controller through a compiler which creates the REIS RRL robot language code from the IRL file. All robot positions and welding process parameters are encapsulated in the final program.

GRASP Simulation Model of the REIS Robot Cell

6.2.5 The Laser Scanner

To plan the welding, a laser scanner, mounted on the robot, measures the seam-profile by weaving a laser-beam across the seam. While the seam is being measured, the robot motion is controlled by a PC. The automatic measuring system has been developed in collaboration with Aalborg University in Denmark within the framework of a Danish research program called IPS - Integrated Production Systems.

Fig. 6.10. The Laser Scanner Mounted on the Reis Robot Arm.

6.2.6 The InterRob Database

The InterRob database is an information system which is meant to improve the communication inside the field of industrial robot applications. The system allows unified access to the product information for robotics and automatic programming generation facilities for a specific robot application. The system is optimised for multi-pass pipe welding, but it is possible to use this system for other applications. The core of the system is the object-oriented database management system, Object Store and STEP TOOLS. The system consists of a product model database, a process database, an equipment database, a file manager database and a production data database. The above systems are put together by a graphically MOTIF based user-friendly interface.

In the database, data necessary for the pipe welding, e.g., data from the laser scanner, product model geometry, the robot model, etc. is stored and through the database all data used in the applications is handled.

6.2.7 Description of the Demonstration

A second, final demonstration was performed at Odense Steel Shipyard (November 1995) in the test laboratory of the Research and Development department. The first prototype demonstration was held in October 1994. The second demonstration consisted of:

Database Demonstration:

The idea of this demonstration was to give an overview of the InterRob database and to show the use of the database for the pipe welding application. This means:

- Creation of a database with import and export of a STEP pipe file
- Conversion of parametric HICADEC pipe file format to B-Spline surface
- Calculation of groove geometry including scanning with laser scanner
- Importing the groove geometry adjustments and modifying the groove geometry
- Calculation of the weld orientations and weld points along the seams
- Generation of the RRL robot program and transfer for dry-run execution on the robot.

All necessary calculations and handling of data necessary for programming the robot has been included. This matches the yard's needs for fully automatic off-line programming from CAD geometry.

GRASP Simulation and Off-line Programming Demonstration:

The GRASP system is placed at the centre of the data flow from CAD/Database to the actual robot. The demonstration illustrates the following points:

- Import of the pipe geometry as NURB curves and surfaces via STEP
- Import of the weld seam geometry via STEP which is derived from the laser scanner data to determine the corrected curve due to pipe distortion
- Import of the welding process parameters associated with the multipass welding task
- Semi-automatic off-line program generation in GRASP for the welding task
- Animated simulation and verification of the welding task in GRASP, identifying clashes, singularities and other robot constraint violations
- Output of the simulated robot program to the REIS robot via the neutral language IRL and then the REIS language, RRL. All welding process parameters are kept intact from the source CAD/Database including weaving parameters
- A 'dry run' (without the welding torch switched on) will be shown on the real robot, start-point sensing included
- Improvements to the REIS robot accuracy (to less than 1.0 mm) will be illustrated by discussing and demonstrating the calibration procedures developed by BYG in GRASP. These procedures are the same as those used for the Rolls Royce demonstration.

Demonstration of Dynamic Robot Simulation:

This demonstration was connected to the work performed in Work Packages 1 and 2. The Dynamic simulation performed in the ROPSIM system (Fig. 6.11) was also shown on a Reis RV6 robot, using one of the robot applications at the shipyard.

Fig. 6.11. The Reis RV6 Robot used for the Demonstration of Dynamic Simulation

At the demonstration the robot was be programmed to follow the standard Schmid curve drawing the curve with a pen. The results from the program execution could then be compared to the simulation results. Both, the robot and the simulation system were given the same path plan in IRL. Further details on the simulation system are given in Annex 3.

Dynamic simulation is not exploited at the shipyard. Present use of robots are mainly for arc welding. The speed of the robot is in this way limited to the process speed and dynamic simulation is unnecessary. In future where painting and assembling applications will be likely, the robots will be programmed with much higher speeds, thus increasing the need for dynamic simulations.

Pipe Welding:

Finally, the demonstration showed a complete welding of a connection pipe. The welding consisted of one root layer welded in short arc. The welding was performed on a pipe with changing root gap and the results from the welding correspond to the quality demands from production. After this layer, two more seams were welded to fill the groove. These two seams were be welded with a much higher speed and in spray arc.

6.2.8 Benefits and Commercial Impact of Results

In the highly competitive shipbuilding industry, price and time to market are essential for competitiveness. On the other hand, the rapidly changing demands from ship owners concerning new vessel types, sizes and performances must be considered. For advanced vessels like container vessels and fire fighting tugs extensive use of subcontractors might be considered, while other ship types demand more manpower from the shipyard. The result is a need for a highly flexible production. This has great impact on the layout of automation and for the use of robots.

Usual on-line programming of production equipment is far too time-consuming. At the present time all robot applications running at the yard are programmed using CAD geometry and automatic off-line programming. Advanced simulations create the basis for developments of new applications and for testing of off-line generated production programs. The system architecture is based on participation in seven ESPRIT projects during the last ten years.

Odense Steel Shipyard has integrated InterRob database applications for handling of product models and production data into an application combining information technology with process knowledge and modelling techniques. This combination has made it possible for the shipyard to enter one of the most difficult areas of automation.

6.2.9 Future Application of Results

After the conclusion of ESPRIT InterRob the pipe welding application will be moved to production. After a few changes it will be used for the production of pipes. Another Reis robot, having the same controller features will shortly after be used for producing T-profile components.

Simulation of robotics tasks is already widely used in the company. Simulation is used to prepare welding tasks and for development of automatic off-line programming methods. The use of STEP has shown to be a powerful tool.

Future applications will not only be applied to welding and cutting but also painting, assembly, and other processes will be considered. The need for different robots and related equipment will simultaneously increase and so will the need for standardised communication. It is obvious that experience gained from InterRob will be used to form these future applications.

7 Contributions to Standardisation Activities

A. Ludwig[1], T. Horsch[2]
[1]Forschungszentrum Karlsruhe GmbH Technik und Umwelt, IAI
P.O. Box 36 40, D-76021 Karlsruhe, Germany
[2]Reis Robotics
P.O. Box 11 01 61, D-63777 Obernburg, Germany

7.1 InterRob and the STEP Standardisation Work

STEP, the unofficial acronym for the International Standard ISO 10303, means *Standard for the Exchange of Product Model Data*. The official title of this standard, which consists of many parts, is *Industrial automation systems and integration – Product data representation and exchange*. STEP is being developed by an international community of experts within ISO TC184/SC4 since 1984. Europe has been actively involved in this development, at least until the early nineties, mainly via the contributions of several ESPRIT projects. The first one of these projects which had significant influence on STEP in the time period from 1985 through 1989 was ESPRIT 322, CAD*I (CAD-Interfaces). Project ESPRIT 2614/5109, NIRO (Neutral Interfaces for Robotics), continued the contributions of CAD*I and added a new aspect: kinematics. The InterRob Project, partially in cooperation with ESPRIT Project 6040, PRODEX (Product Model Exchange Using STEP), carried on this work with respect to application documents for the design using surface models, to the implementation specifications, and, last but not least, to the completion of the kinematics integrated resource model.

The direct participation of the European industry in the development of STEP has increased since 1990; an example is the predominant role of the industry in the development of the Application Protocols AP 214 (Automotive Design) and AP 212 (Electrotechnical Design) which have been launched since 1991/92 by the German ProSTEP initiative. But also these projects were significantly based on the experience gained in the ESPRIT project teams.

At the beginning of the InterRob project those parts of STEP which belong to the so-called 'Initial Release' had just passed their Committee Draft (CD) ballots. A CD ballot is the first 'official' voting on the draft of an International Standard which precedes its publication as a 'Draft International Standard' (DIS). The Initial Release collects those parts that are considered essential for any application of the standard, and additionally two Application Protocols (AP) which should allow to actually start implementations:

- Part 1: Overview and fundamental principles
- Part 11: EXPRESS language reference manual
- Part 21: Implementation methods: Clear text encoding of the exchange structure
- Part 31: Conformance testing methodology and framework: General concepts
- Part 41: Integrated generic resources: Fundamentals of product description and support
- Part 42: Integrated generic resources: Geometric and topological representation
- Part 43: Integrated generic resources: Representation structures
- Part 44: Integrated generic resources: Product structure configuration
- Part 46: Integrated generic resources: Visual presentation
- Part 101: Integrated application resources: Draughting
- Part 201: Application protocol: Explicit draughting
- Part 203: Application protocol: Configuration controlled 3D designs of mechanical parts and assemblies

Hence, the main effort in SC4 during the first year of InterRob was directed to complete the Initial Release parts. Their DIS versions were distributed in the mid of 1993, and they were submitted to ISO for publication as International Standards (IS) at the end of 1994; the actual publication took place in the first months of 1995. The work on all other parts in the development chain suffered an unpredictably long delay until the Initial Release had reached DIS status. This holds especially for all that work that required 'external' services (i.e., from outside the project team that performed the technical development), like Qualification and Integration review by the official groups in ISO TC184/SC4. Although these services were bottlenecks also after that time, the development could be accelerated about at the end of the first project year of InterRob.

In the course of the InterRob project, project members were actively involved in the standardisation process for STEP. The main areas of contributions from InterRob to STEP were:

- the development of the kinematics information model for STEP in ISO 10303-105;
- the final definition of the physical file format in ISO 10303-21, which belongs to the Initial Release;
- contributions to the development of the *STEP Data Access Interface* (SDAI) in ISO 10303-22;
- the development of the STEP Application Protocol for mechanical design using surface representation, ISO 10303-205, and of related AIC documents. AIC means *Application Interpreted Construct*, which is a portion of an Application Protocol that is shared by at least one other AP.
- the InterRob Specification of a STEP Based Reference Model for Exchange of Robotics Models.

The main InterRob contributors to STEP were Forschungszentrum Karlsruhe and SINTEF. H.-P. Lorenz and A. Ludwig from Forschungszentrum Karlsruhe acted, one after the other, as Project Leader and Part Editor for the STEP kinematics model. J. Haenisch from SINTEF has held these positions for the activities related to AP 205. E.G. Schlechtendahl from Forschungszentrum Karlsruhe has been, among other activities, an active member of the STEP implementation Working Group.

7.1.1 The STEP Kinematics Model

InterRob has inherited from NIRO the Project Leadership and Part Ownership for the STEP kinematics model, ISO 10303-105, which is an 'Integrated application resource' part of STEP. At the beginning of InterRob, a Working Draft was available: Its preparation for CD ballot had been delayed for a long time due to the priority of the Initial Release. The technical content of the CD version, sent out for ballot in December 1993, has undergone very little change as compared to the preceding versions, but the internal structure of the information model was changed drastically as a consequence of the STEP integration process. A typical example for these changes was already given in Sect. 4.1.1 of [Bey 1994].

After the CD ballot, in May 1994, the preparation of the DIS version begun. Due to some ballot comments (primarily launched by InterRob), the list of pair types has been enhanced from 12 to 16 types, some definitions were made more precise, and some minor technical improvements were incorporated. The main effort, however, was again caused by requests of the STEP Integration, Qualification, and Editing reviews. In the course of these processes, the quality of the document has admittedly been improved, and in spite of some necessary concessions to the information modelling style imposed by the Integration team, a functionality could be restored that had been lost (or, at least, hidden) in the CD version, namely, to mount a mechanism not only on the ground, but also on a link of another mechanism. Besides this, the separation of kinematics information and shape information was performed even more stringently, with an associating entity allowing to associate zero, one, or more shape representations with a single link representation. Some other changes were induced by modifications in Parts 41, 42, and 43 which occurred in the course of their promotion from DIS to IS level, and led to a more well-structured embedding of the kinematics model as a product property definition into the overall product structure model of STEP.

The DIS version of ISO 10303-105 has been submitted to the SC4 secretariat in December 1994; the DIS ballot, which ended in late December 1995, turned out to be very favourable, since almost all votes indicated approval (with some comments being attached). This allowed the IS version of ISO 10303-105 to be prepared, even though after the actual InterRob project period. It has been submitted to ISO for its publication as an International Standard in June 1996.

Work Package 1 of InterRob had decided to rely on the IS versions of the Integrated Resources of STEP for the product structure and shape models of the InterRob specification. As an inevitable consequence, Work Package 1 had also to upgrade the software implementations from the CD version to the DIS version of Part 105 for the kinematics model, for sake of consistency. Since both, the IS versions of the Initial Release parts and the DIS version of Part 105, were not available before late 1994, this implied some updating of the InterRob STEP processors during the last project year. However, this additional burden is balanced by the advantage to have processors which are up to date and comply to the actual standard as far as possible.

7.1.2 Physical File Structure

STEP Part 21 belongs to the Initial Release, and one should think that there has been little work left after the DIS version had been released for balloting. However, the DIS ballot evaluation has induced a lot of more or less significant changes to almost all Parts of the Initial Release, including Part 11, the EXPRESS language definition, and Part 21. Thus not only the ballot comments on Part 21 proper, but also any changes especially of the EXPRESS language had to be reflected when preparing the IS version of ISO 10303-21.

The most important issue in this context has been the mapping of entity attributes that are select data types to the physical file structure. An EXPRESS select data type defines a list of data types, called the 'select-list', whose values are valid instances of the select data type. The data types in the select-list may be either entity data types or other defined data types. In the case of an entity data type, there is no problem, because the entity instance has the entity name as its keyword, and thus carries the type information with itself. This had originally not been provided for defined data types in the select-list. In some circumstances, the underlying basic data type may be the same for several of the select items. For example, *measure_value* is a select data type defined in Part 41 with many select items, like *length_measure*, *time_measure*, *plane_angle_measure*, and others. Most of these items are defined as to finally evaluate to REAL values. Thus the information about what kind of *measure_value* is actually encoded would be lost in the exchange structure, if just this underlying data type would be encoded.

Therefore, the solution agreed upon for the IS version required to encode an attribute which is defined by a select type by giving the entire 'path' from the given select type to the actually chosen leaf of the select tree. This solution has been implemented in the InterRob STEP processors. Unfortunately, an intermediate version of the Part 21 document, which was not fully self-consistent, was submitted to ISO for publication by a mistake. This forced the SC4 community to prepare a *Technical Corrigendum* to the IS version, which has been approved by SC4 in October 1995; here, the mapping rules for the select type have again been modified – actually, they have been relaxed to some extent. Of course, these new mapping rules could not be reflected anymore in the InterRob implementations.

7.1.3 STEP Data Access Interface

SDAI is the acronym for *STEP Data Access Interface*, specified in ISO 10303-22. SDAI shall become the specification standard for accessing STEP-structured data either in data bases or in primary memory from application programs. The interface is specified as a structured set of service routines (functions) in a way that does not depend on a particular programming language. Specific programming language implementations of this generic SDAI will be specified in separate standard documents as so-called language bindings. At the end of 1995, the most advanced language binding document is ISO 10303-23 (C++), followed by ISO 10303-24 (C) and ISO 10303-26 (IDL). The project for developing a FORTRAN binding specification is being cancelled at this instance of time.

Like in NIRO (see [Bey 1994]), InterRob members participated in the SDAI specification work. One of the main issues were the treatment of transactions. Other major issues were concerned with the appropriate definition of data repositories, with the visibility of instances from referenced schemas and, more generally, with the relationship between different schema instances.

ISO 10303-22 passed a second CD ballot successfully in early 1995; the DIS version is to be submitted to ISO for balloting at the end of 1995. If no further delays are encountered, ISO 10303-22 may become an International Standard at the end of 1996. The language binding documents will probably follow.

7.1.4 STEP Applications of Surface Modelling

A Working Draft of AP 205 *Mechanical design using surface representation* had already been developed in the framework of ESPRIT project 2195, CADEX (CAD Geometry Data Exchange), and, hence, was available at the beginning of the InterRob project. Major parts of its contents have been incorporated into AP 203 and, in this way, became part of the Initial Release. The further development of the proper AP 205, however, suffered from the priority given to the Initial Release, too. In the meantime, the document has been owned by an InterRob member, who also acted as the Project Leader; AP 205 successfully passed its CD ballot in the spring of 1995. In the fall of 1995, its state is officially announced as 'preparing for DIS ballot', but, unfortunately, the positions of Part Owner and Project Leader are vacant, and the AP 205 project is actually on hold.

This does not necessarily make all the efforts obsolete which were dedicated to the former development. Already the fact that AP 203 is sharing some constructs with AP 205 launched the development of a couple of AIC documents:

- AIC 507: Geometrically bounded surface
- AIC 508: Non-manifold surface
- AIC 509: Manifold surface
- AIC 517: Mechanical design geometric presentation
- AIC 518: Mechanical design shaded presentation

All these documents, which passed successfully a combined New Work Item and CD ballot at end of October 1995, were edited and also otherwise substantially influenced by InterRob members. Their importance has been raised by the fact that some other APs that are on different development stages do also need surface models, e.g., AP 214 (Automotive design), but also APs for the shipbuilding area, like AP 216 (Ship moulded forms); due to this overall interest they will be completed, regardless of the fate of AP 205.

7.1.5 The InterRob Specification

This last contribution has been described in more detail in Sect. 3.2. The InterRob specification may be regarded as the Application Interpreted Model (AIM) part of an Application Protocol for robotics applications. There are three main differences to STEP APs: First, the Application Reference Model (ARM) is not included in the specification (it has been a separate deliverable [Sørensen 1993]), and there is no formal mapping from the ARM to the AIM. Second, the references to STEP resources are expanded explicitly in order to have a self-contained specification; an actual AP would leave many such references unresolved. Third, there are many constructs, particularly in the subschemas for robotics, dynamics, control, and calibration, that cannot be found in the Integrated Resource parts of STEP; according to the AP development rules established in ISO TC184/SC4, these constructs should be defined in additional new Integrated Resource parts in order to use them in an actual AP.

The InterRob specification has been informally notified to the ISO community, but was not formally submitted, as this was not intended, because of the lack of formal compliance with AP rules, and since not enough support could be found up to now to launch a New Work Item (where at least five countries must commit themselves to participate actively). The main purpose of the specification was to serve as a basis for implementing the InterRob STEP processors. Nonetheless, there was a mutual influence between the work on actual STEP documents and on the InterRob specification: The experience gained from STEP facilitated the task of making the InterRob specification a 'STEP-like' document, and the experience gained in writing this specification and in implementing it was used in updating the various STEP models, e.g., the Kinematics model in ISO 10303-105.

7.2   InterRob and the IRL Standardisation Work

This section gives a brief overview about IRL and describes the contributions to standardisation activities mainly to the German standardisation board DIN/NAM AA 81.4 and the status of the standardisation on international level.

IRL is an abbreviation of Industrial Robot programming Language and defines a modular, compiler oriented, PASCAL-like language, which offers general functionality of a high level language. The main characteristics of this language are additional features, which are robot related including geometric data types and operators, move instructions, interaction with external devices (gripper, external axes) and parallel execution of programs.

The main aspect of this language is to provide a neutral robot programming language to enable end users to be independent of any particular robot controller. Nevertheless IRL is presently not implemented as a native language in any commercial robot controller, but interface processors (translators) to and from established robot languages like the **Reis Robot Language RRL** have been developed. The IRL processors developed in InterRob are llisted in Annex 4.

InterRob has made significant contributions to the standardisation of IRL. These contributions include statements for sensor integration, parallel processes, multitasking, control of the welding source, weaving parameters etc. Many of these suggestions hasve successfully been introduced to the working group of IRL.

The definition of IRL has been a valid German national standard (DIN 66312, IRL Part 1 since June 1993 and IRL part 2 since 1995). The situation on the international level is different. At the ISO/TC184/SC2/WG4 meeting in Budapest in May 1994 item 11513 "IRL/PLR" was deleted from the action list. Even more at the CEN/TC310 meeting in Copenhagen, which was held in June 1994, the existing German standard DIN 66312 was rejected. The reasons for this conflict of interest between different standardisation comities may be, that on the one hand robot vendors do not see a competitive strength to implement a neutral language and on the other hand the end users do not in general have flexible production which requires such a language. Off-line programming is still a front end technology with relatively few users outside the automotive industry.

The international working group ISO/TC184/SC2/WG4 is currently discussing to continue work on graphical user interfaces for robot programming and operating.

Independent from the standardisation work the automotive industry has started the project Realistic Robot Simulation (RRS) together with several robot vendors and suppliers of CAR (Computer Aided Robotics) tools to improve the simulation accuracy for downloading programs and executing them without a large amount of time spent for re-teaching.

8 InterRob Project Results and Exploitation Plans

K. Hasund et al.
SINTEF Informatics
P.O. Box 124 Blindern, N-0314 Oslo, Norway

8.1   General

The aim of the InterRob project is to demonstrate precision manufacturing with off-line programmed robots, using electronic product data technology. InterRob will show the feasibility of a complete chain of open manufacturing systems, integrated through the use of standards in the CAD/CAM and robotics and the value of such integrated manufacturing systems to user companies. Standardised interfaces allow the exchange of components within such an integrated system If the single components of such an integrated information system (e.g. CAD, simulation, programming and the robot systems) to be made without disturbing the whole system, in addition standardisation enables efficient exchange of information between companies.

Open systems are not only of interest to user companies, but also to small vendors, who can only supply parts of an information chain. Large corporations are using this technology to improve both their internal communication and their information exchange with sub-contractors.

Currently the use of product data technology and standards is in its infancy in most companies.

Most vendors and users are waiting for the standards to stabilise. The vendors tend to wait until a market emerges and the users wait for a stable product that conforms to a recognised standard. The contribution of InterRob to this scenario is schematised in Fig. 8.1:

1. The vendors (BYG and REIS) provide standard interfaces to their products, making them fit into an open integrated information chain thus increasing the availability of products in the "open market".
2. The users (Rolls-Royce and Odense Steel Shipyard) have experience of open systems in their information chain and demonstrate the feasibility of operating open standards in a realistic industrial environment.

3. The research institutes (DTU, FZK and SINTEF) acquire technology and provide technology support to both vendors and users making both the concepts and their implementation possible.

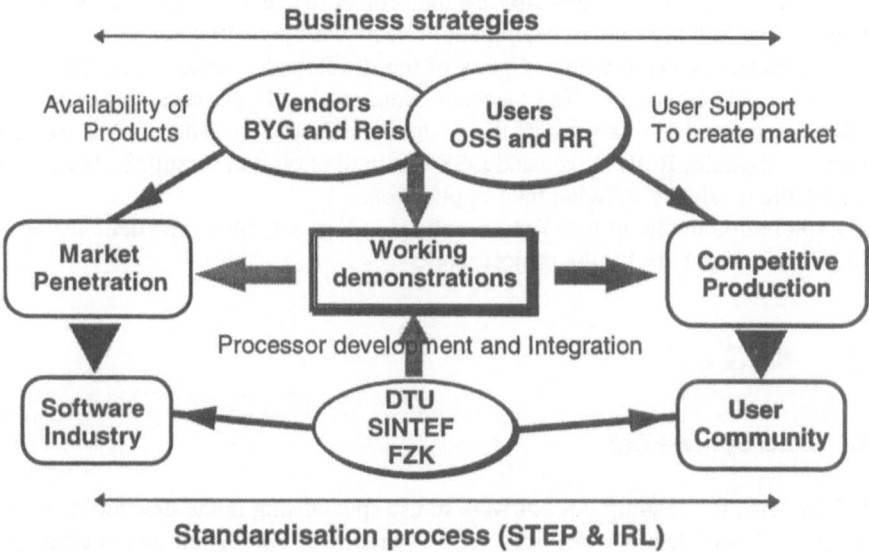

Fig. 8.1. The InterRob Exploitation process

Successful demonstrators create interest from other users, in this way a market for open standards will be created. It is our hope that through this project we will inspire more vendors to offer standard compliant systems and consequently more users to implement this technology.

The InterRob consortium sees its main task in exploitation as the enlargement and enhancement of the standards STEP and IRL and the exposure of the results to industry across Europe.

Primarily exploitation of InterRob results means convincing other companies that standardised open systems have a significant advantage over existing ones; more precise and flexible manufacturing, more efficient information exchange and increased flexibility in system configuration. The current slow growth in the use of standards like STEP for geometric data shows it is important that industrial interest (and pressure) groups, like the InterRob consortium promote such ideas.

The vendors in the InterRob Project (BYG and Reis) will assure the availability of developed open systems at the end of the Project. However, these companies are not big enough to create a market for open systems, they need the support of the user community to convince CAD and other simulation and robot system vendors to commit themselves to STEP and IRL.

The processor development and integration work done by DTU, FZK and SINTEF is not primarily intended to earn money but to demonstrate the InterRob solutions works.

These research organisations will use the project results as a baseline for further research work and in future development projects with industrial companies.

The commercial exploitation of parts of the developed software is an important by-product of InterRob. The system vendors will actively promote the implemented standards as parts of their commercial systems. The same is true for the user companies Rolls-Royce and OSS who will exploit the results in-house and promote the results by showing their applications.

The exploitation model in InterRob visualised in Fig. 8.1, shows the demonstrators as the centre of gravity for the project work.

8.2   BYG Ltd

8.2.1 BYG Systems Ltd

BYG Systems is a leading UK software house specialising in the development and provision of products and services for 3D graphical simulation and multimedia authoring. Within the simulation field BYG has along history and considerable expertise in the area of robotics modelling and off-line programming. It has been successful in winning several major European funded projects such as NIRO (Neutral Interfaces for Robotics, number 2614/5109, 1989-1992) related to robotics data exchange standards. It is therefore committed and has already invested in and contributed to the development of new product data technology in the robotics arena.

BYG's involvement in the InterRob project is a natural progression from the NIRO project, further enhancing its position in the market-place for pre-competitive development of software interfaces and functionality in robotics standards. Not only does it mean that BYG can benefit from earlier products to market it also has the opportunity to influence the direction in which the standards development progresses.

The InterRob project is not simply about robotics standards per-se. It also addresses the more global issue of improvements to accuracy, usability and productivity in CIME. This means that certain practical issues such as robot calibration and better off-line programming methods are also addressed. The project must demonstrate the feasibility of the standards and processors being developed in an industrial environment. This means that BYG can benefit greatly from the development of real (prototype) improvements to its off-line programming and simulation tool in an industrially proven (i.e. useful) context.

This is very much in line with BYG's aim to provide software tools that meet the demands of the end-user whilst offering the very latest in technological advances.

8.2.2 InterRob Results

STEP and Product Modelling. The investment in time and capital expenditure by companies using CAD/CAM can be considerable. Without standards in product modelling (STEP) the introduction of new technologies and methodologies such as 3D graphical simulation and off-line programming would normally require re-modelling of CAD data etc. As other CAD/CAM suppliers move towards STEP in the future it will be vital for BYG to be able to support the emerging standards. The creation of prototype STEP processors for GRASP is a great advantage for BYG in its sales and marketing efforts in the world marketplace. Specifically, the expertise in the STEP arena will allow BYG to enhance its ability to provide a standard data exchange mechanism with other compliant software tools, including other simulation software. The inclusion of NURB (Non Uniform Rational B-Splines) curves and surfaces in GRASP via STEP import greatly enhances the generality of the software and makes it more widely applicable to many more users in different areas of interest.

Off-line Programming. Other important specific results that will be exploited by BYG include the prototype developments of the GRASP application tool for off-line programming in arc welding and plasma spraying.

The high level menu interface in GRASP for arc welding makes the generation, verification and optimisation of robot welding programs off-line much easier. The functions in the arc welding menu are fully integrated with the STEP product model meaning that technological data such as torch orientation and process data including arc power, weaving patterns etc. are closely associated with the weld geometry even before the user intervenes to complete the creation of the robot program. The flexibility and ease of use offered by this integrated approach makes the tool very productive and a valuable asset to the creation and maintenance of robot programs off-line.

A menu in GRASP for the creation of robot programs for plasma spraying is another useful development from the InterRob project. It has clear potential for use in robotics paint spraying applications, a very common activity for robot installations in industry. With relatively little effort BYG's existing functionality for the plasma spraying functions could be extended for paint spraying. BYG will capitalise on this new feature in the market-place.

Robot Accuracy - Calibration. It is an accepted fact that off-line programming of robots is beset by problems arising from their inherent inaccuracy. The advances made by BYG in its integrated calibration system for GRASP have provided a cheap, easy to use tool for improving the accuracy of off-line produced robot programs down to less than 1 mm. This has been done in a novel way by using the robot itself as the measuring device, without the need for expensive or bulky equipment. BYG will be able to offer this tool to its users as part of the integrated off-line programming system in GRASP.

Contribution to Standards. BYG's GRASP product is positioned in between the end point (the robot) and the originator (the CAD system) in the production process. It therefore requires the ability to transfer data via STEP for the product model data and IRL for the robot programs. BYG has a deep knowledge and experience in robot modelling, programming, simulation and specific robot processes such as arc welding. It has therefore contributed significantly to many areas in the development of the STEP and IRL standards in the InterRob project. Specifically this has included:

- STEP product model data schema for weld geometry (weld seams)
- STEP product model data schema for weld process parameters
- STEP product model data schema for calibration data
- STEP product model data schema for robots with auxiliary axes
- IRL program definitions, especially for robot control (configuration rules) and process instructions for welding.

8.2.3 Exploitation Achievements

BYG has promoted the InterRob project results on several levels throughout its lifetime and will continue to do so in the future since it has a vested interest. Specific activities carried out during the project include:

- A seminar on STEP held at BYG Systems Ltd in November 1995. Open to industry and research/education this seminar had speakers from Rolls Royce, Intergraph, CADDETC, Loughborough University and BYG.
- A multimedia CD-ROM application was developed by BYG using its own TX-Authoring software tool to disseminate the results of the InterRob project. This was distributed to the Commission and within the Project Consortium (50 copies) itself. This type of dissemination is an accessible and attractive way of communicating project results to a wide audience.
- Five liceDnces of the GRASP software have been sold during the course of the InterRob project which have been made possible in part because of the prototype results emerging from it. The points of interest relating to the InterRob project relevant to these sales was on the issue of calibration (i.e. robot accuracy) and the arc welding menu.

In February 1994 BYG secured the decision to use the results from the InterRob project in the STEP standard in another European Project. The Brite-Euram project PROMET (Rapid Prototyping of Metal Components) needed to transfer

CAD data about the parts to be produced by the robotic weld deposition process which also had to include the torch orientation, the weld process parameters and the torch orientation. The InterRob project results are being directly used in this BRITE project.

8.2.4 Future Exploitation

BYG will focus on bringing its prototype products (STEP processors, Arc Welding and Plasma/Paint Spraying applications for GRASP) to market as part of its continual commercial sales and marketing efforts. To BYG's mind it is clear that these developments address the needs of the end user, proven by the involvement of major industrial collaborators such as Odense Steel Shipyard and Rolls Royce.

This will be achieved using BYG's international sales and marketing network both in Europe and the Far East.

8.3   DTU

8.3.1 Main Business of DTU

The Technical University of Denmark (DTU) is a government owned University. The general purpose of DTU is to educate engineers at M.Sc. and Ph.D. levels, carry out technical-scientific research, and participate in the technical developments.

8.3.2 General DTU exploitation achievements

Besides geometrical and kinematic data, STEP product models have been extended to include product definition data for robotics, and for manipulator and controller dynamics. This facilitates more realistic and accurate simulation of robotic systems and allows for import and export of product models in simulation.

On the basis of the InterRob Application Protocol (schema) the following processors are developed and tested:

- STEP pre-processor to CATIA
- STEP post- and pre-processor to ROPSIM

On the basis of the IRL definition the following processors have been developed and tested:

- IRL post-processors to ARLA and ROPSIM
- PC-based IRL post-processors to ABB-IRB2000.

The above described expertise and software developed in InterRob are intended to be further exploited in course work, continuing research, and industrial consulting activities at university level.

8.3.3 Contribution to Standardisation

Together with other partners DTU has contributed to the ISO document ISO-10305-105 Kinematics and has given inputs to STEP enhancement projects towards the new topical areas manipulator dynamics, controllers, and robotics. Further the InterRob Application Protocol specifying a STEP-based reference model for exchanging more complete robot models has been worked out and sent to ISO TC184/SC4/WG3. DTU has also participated in the formulation and testing of IRL/ICR proposals sent to ISO.

8.3.4 Contribution to dissemination by DTU

DTU has given presentations on InterRob results in several seminars within the Danish national research programme, Integrated Production Systems, IPS, and within seminars for industrialists. Further papers have been presented at:

1. IFIP workshop on Interfaces (1993)

2. Advanced NATO Research Workshop Integration: Information and Collaboration (1993)
3. ESPRIT-CIME 10th Annual Conference (1994)
4. CAPE '95 (1995)

8.3.5 Why InterRob is relevant for DTU

In the InterRob project the open systems' concept is used to identify and generate the information flow via neutral interfaces from CAD via robot task planning and programming to real-time execution. The derived information modelling and model exchange of product definition data, between different systems are precisely the fundamental background needed for teaching CIME-robotics courses at a university level and for initiating further research.

8.3.6 Use of InterRob results at DTU

The InterRob results have been relevant for (1) teaching robotic courses at university level, (2) for several M.Sc. and Ph.D. theses within CIM and integration of robotic systems, (3) the development of a new robot programming and simulation system using neutral interfaces for import and export of models, (4) and for several industrial consulting activities.

8.3.7 Value of contributions/achievements so far at DTU

A major contribution/achievement is the development of an InterRob Application Protocol. The protocol includes sub-sets of ISO-10303 standard parts for product description, supporting a class of solid models denoted polyhedrons or faceted-boundary representations, and face-based surface models, as well as sub-sets of ISO-10303 draft standard for kinematics. The surface models also feature Non-Uniform Rational B-Splines (NURBS). Further the protocol features information in EXPRESS for dynamics and control design, and robotics information, with the intention to cover the functionality of present mechanical product systems and to identify the requirements for future CAD and robot-simulation systems.

8.3.8 Planned use of InterRob results, role in future strategy and products at DTU

InterRob results will be used in continuing efforts within teaching and research in integrated production systems. Licenses on developed software for the exchange of product definition data will be granted to other, external parties. DTU will exploit pre-processors from CATIA to STEP, and processors for IRL to ARLA (ABB-IRB 2000) and IRL to ROPSIM (e.g. license agreements, sales).

8.4   Forschungszentrum Karlsruhe

The Initial Release parts of STEP became International Standards at the end of 1994. As a consequence, the transfer of product definition data becomes more and more important for the industry, and several CAD system vendors are offering corresponding STEP processors. Also processors for the data transfer to KISMET, based on this new standard, have been requested by some users. The STEP–KISMET processor, which has been developed by the Forschungszentrum Karlsruhe in the framework of its InterRob work, has been delivered to JET (Joint European Torus, Culham, UK) in November 1995 and is used by JET for the transfer of CATIA models to KISMET. Its delivery to the commercial KISMET distributor is subject of ongoing talks.

The development of processors for ADAMS, based on the InterRob specification, enables the re-use of product data (kinematics, dynamics) which have been generated by other simulation systems, e.g., KISMET, and vice versa.

The knowledge about the schema driven development of object-oriented software, as performed in Work Package 3 of InterRob, is being disseminated in lectures about Software Engineering to students at the Berufsakademie Karlsruhe.

InterRob has contributed substantially to the development of the STEP Kinematics model, ISO/DIS 10303-105. which has been registered as a Draft International Standard about half a year before the end of the project's period. During all the project time, the project leader and part editor for this STEP part within the ISO subcommittee 'Industrial Data', ISO TC184/SC4, has been an InterRob member from Forschungszentrum Karlsruhe. Furthermore, the Implementation parts of STEP, ISO 10303-21 ('Clear Text Encoding', part of the Initial Release) and ISO 10303-22 ('STEP Data Access Interface', being promoted to the Draft International Standard level at the end of 1995), have been influenced significantly by an InterRob member from Forschungszentrum Karlsruhe. Thus an outstanding expertise has been gained in STEP technology and, more generally, in standardisation procedures.

The contents of ISO/DIS 10303-105 have been discussed in detail during several meetings with experts from the German automotive industry, and contributions have been made to the development of the STEP Application Protocol ISO 10303-214 ('Automotive Design') which has reached the status of a Committee Draft (CD) in August 1995.

Generally, all the experience gained in the InterRob work will be re-used in upcoming new projects.

8.5   Odense Steel Shipyard (OSS)

At Odense steel shipyard, the implementation of robot technology for automation was based on research in several ESPRIT projects undertaken during the last ten years.

The pipe welding application in ESPRIT InterRob will after project conclusion be installed in production. The application will be used for welding of highly complex geometrical shapes, and is as such the most noticeable result from InterRob.

Project results are in addition applied to database developments in the company. Odense Steel Shipyard is developing databases for handling of production facilities and product models, and experience gained through InterRob will be used to implement STEP standards for these databases.

Except from implementation of project results directly in production applications, developments in InterRob are also combined with developments in the ESPRIT project CLEOPATRA (smart welder).

Project results will also be used as a basis for the OSS work in the proposed SEASPRITE project concerning standards in the shipbuilding industry.

Several fundamental technologies have been explored and developed in a variety of projects. The shipyard strives to crystallise such technologies into generally applicable products which are not only used in production at the shipyard, but also commercially distributed to shipyards in a world wide scale.

8.6   Reis Robotics

The main business of Reis Robotics is to provide turn-key solutions for industrial automation processes with robotic systems. Major applications include robot assisted arc welding, handling, grinding, etc. Exploiting the underlying ideas of InterRob helps Reis to strengthen its market position in the area of computer integrated manufacturing and engineering.

Enabling applications with computer integrated production in (very) small batches without a large amount of time spent for programming will be a key feature of future Reis products. This means that there exist generic interfaces describing the information flow from the design phase to the machine program. Since this involves ideal models of the robot, tool and workpiece, advanced calibration techniques are necessary. Reis has gained the following benefits from the project:

- Better understanding of the various sources of static and dynamic errors
- Development of measurement techniques for improved calibration
- Development of new interpolation algorithms to enhance the model fidelity (NURBS interpolation)
- Improvement of the interoperability of Reis product with IT systems from other vendors in CIME environments

Good absolute positioning and path following accuracy are a prerequisite for robot applications which are generated off-line. Through the developments made in the InterRob project Reis is now able to penetrate new market segments like the automotive and shipyard industry via the improved interoperability possibilities and the quality of its products. A big advantage doing this is the extremely short time to market at Reis. It is planned to show most of the developments done within InterRob at the Hannover industrial trade fair.

8.7   Rolls-Royce plc

The Plasma Spraying Application at Rolls-Royce is an existing process that has directly benefited from the application of InterRob technology. Part program generation has been enhanced with the provision of a Plasma Spray interface and deposition simulation. The time taken to generate and prove robot programs for a new component will be significantly reduced. New components are being identified at present, and the technology can easily be applied to a similar process called Vacuum Plasma Spraying.

An advanced welding process currently undergoing manufacturing development, has only been made viable due to Rolls-Royce's involvement in the InterRob project. Existing Robot processes such as Laser Maskant cutting have also been improved with the application of InterRob calibration procedures.

Rolls-Royce will utilise InterRob technology in new robot applications wherever possible, and as a result of the experience gained as a member of InterRob, will actively seek to participate in future European collaborative projects.

8.8   SINTEF Informatics

8.8.1 SINTEF Group

The SINTEF Group performs contract research and development for industry and the public sector. SINTEF is an independent foundation with its headquarters in Trondheim, Norway. R&D units are located in Trondheim, Oslo, and Mo i Rana. The Group performs projects primarily in technology, but also in the natural sciences, medicine, and the social sciences, for Norwegian and foreign clients. SINTEF is the largest independent research organisation in Scandinavia and has about 2,200 employees, 1450 of these are scientists. The units belonging to the SINTEF group had a total turnover of 1.5 billion Norwegian Kroner in 1994 of which 80% came from contracts for industry and the public sector.

8.8.2 SINTEF Informatics

InterRob was connected to the department SINTEF Informatics in Oslo. The main area of work at SINTEF Informatics are:

- Geometric Modelling
- Industrial mathematics
- Computer supported planning
- Distributed Information Systems

SINTEF is aiming at combining strength from the different technological areas to solve real problems for industry. The work in InterRob is a joint effort between geometric modelling and distributed systems supporting activities in the area of geometry representations, STEP technology and object oriented database systems and architectures.

8.8.3 Objective in InterRob

As an institute working mainly in applied research, SITNEF is heavily dependent on the development of the technological basis as a fundament for research projects with industry. The competitive edge is SINTEF's leading role in selected technological areas.

One of these areas are in Product Data Technology, especially connected to the representation and handling of sculptured surfaces.

A second selected area relevant for InterRob are the activities in integration of and interoperability between heterogeneous systems, mainly based on object oriented modelling, database technology and the use of distributed integration platforms like the Object Request Broker (ORB) concept from OSF.

SINTEF's role in InterRob was to provide technology and expertise in the mentioned areas, especially focused on the areas of standardisation and realisation of the demonstration examples.

To defend the investment in InterRob, results and experience gained in the project are expected to strengthen SINTEF's technical platform and market position in the mentioned areas.

Annex 1 Proposed Extensions to STEP: Selected Parts of the InterRob Application Protocol

U. Kroszynski[1], A. Ludwig[2], T. Sørensen[1]
[1]Danmarks Tekniske Universitet, Instituttet for Styreteknik
Bygning 424, DK-2800 Lyngby, Denmark
[2]Forschungszentrum Karlsruhe Technik und Umwelt, IAI
P.O. Box 36 40, D-76021 Karlsruhe, Germany

This Annex summarizes the InterRob specification by giving the EXPRESS-G diagrams of those parts that go beyond the presently existing parts of ISO 10303 and, so far, represent proposals for extensions to STEP. To the authors` knowledge, these extensions are not in the scope of any project in ISO TC 184/SC4 at the present time.

EXPRESS-G is a graphical subset of the EXPRESS language. Like EXPRESS, also EXPRESS-G is defined in ISO 10303-11 [ISO 1992]

There are four parts of the InterRob specification which are not dealt with here, because they represent subsets of constructs which can be found in the Initial Release parts of STEP or in ISO/DIS 10303-105, the most recent version of the STEP kinematics model. They were the basis to which the extensions were added. In the diagrams of this Annex, the relationships to the original STEP schemas were given, instead of those to the omitted parts of the InterRob specification. The only seeming exception is in Fig. A1.13 where the *ir_g4_schema* and the *ir_k3_schema* is referenced. However, the first schema is actually a subset of an AIC (see Sect. 7.1) which, in turn, is contained in AP 203, and *ir_k3_schema* is the name of the schema for the InterRob Dynamics Information Model.

As usual also in STEP documents, references to basic EXPRESS data types (like REAL, STRING, etc.) have been omitted, in order not to overcrowd the diagrams. Corresponding attribute relationships end with a circle which indicates the direction of the relationship. For the same reason, in the diagrams illustrating the Dynamics Information Model (Annex 1, Chap. 2) also references to any measures have been omitted, independently of whether they are defined in STEP documents or inside the *ir_k3_schema*. It is felt that the given attribute names provide enough information to find which kind of measure is needed.

The Control Information Model is given in Chap. 4 of this Annex in somewhat more detail, because it contains functional modelling.

The InterRob Robotics Information Model

In the Robotics Information Model, some extensions specific to robotics applications are proposed. In Fig. A1.1, there are two specifications of the *linear_path* entity (top) and, more important, the construct for a *kinematic strategy* (bottom) which is needed to control uniquely a robot whose kinematics are redundant.

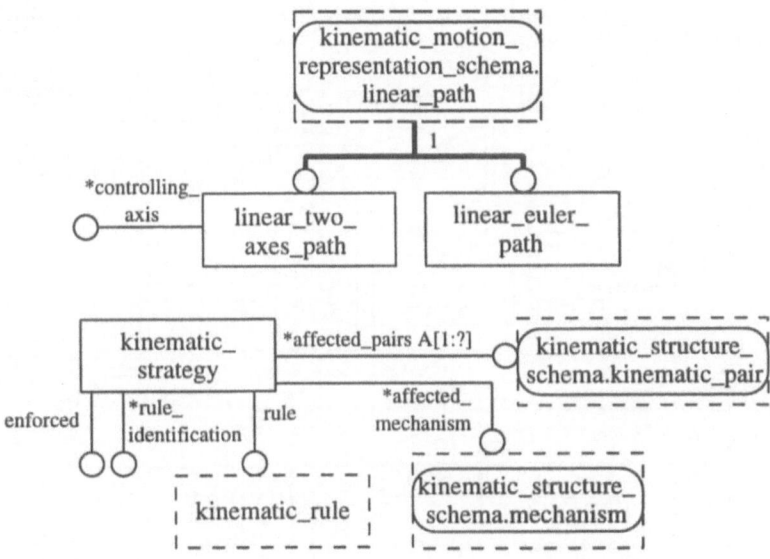

Fig. A1.1. EXPRESS-G diagram of the InterRob Robotics Information Model – proposed extensions to STEP. (Diagram 1 of 3)

Figure A1.2 illustrates the workcell and tool definitions and the means by which potential connections between components inside the workcell are defined. In this context, the concept of workframes shows up; a variety of specialisations is available. The details of the *camera frame* and of the *light frame* are given in Fig. A1.3.

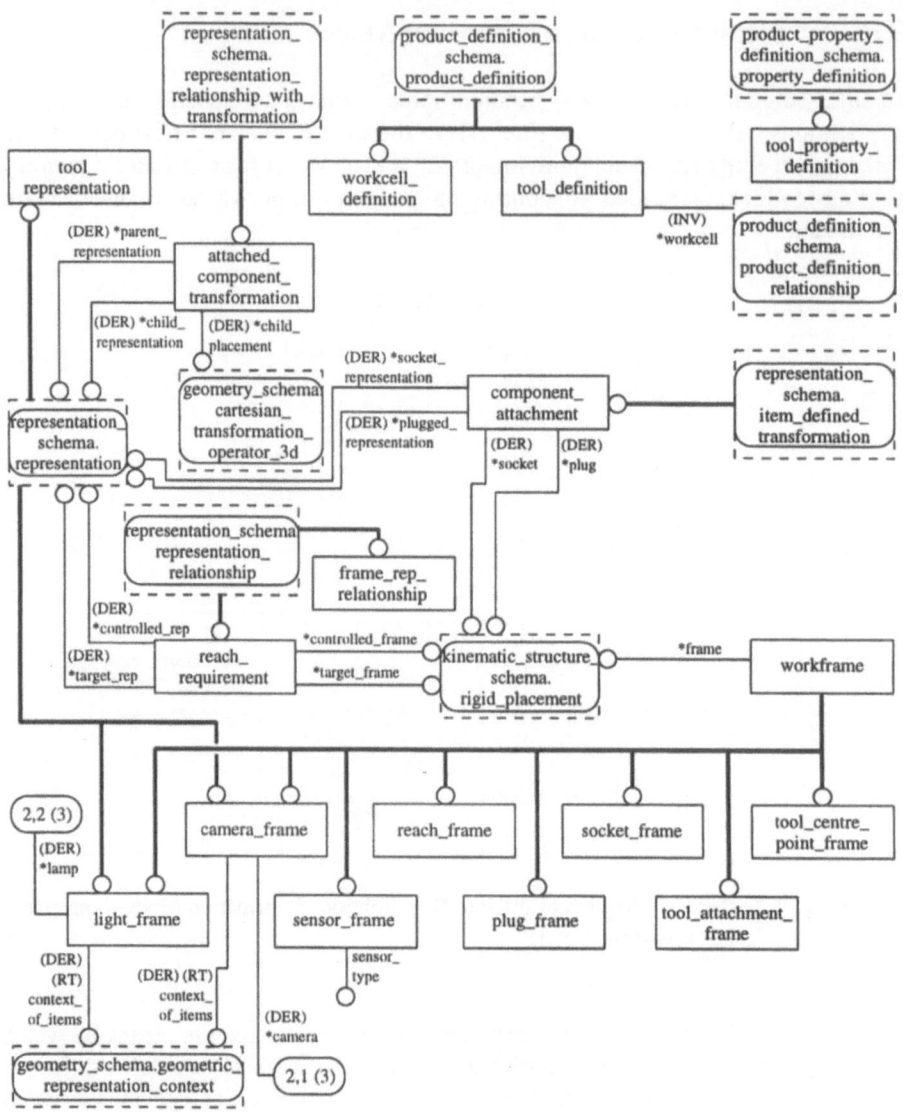

Fig. A1.2. EXPRESS-G diagram of the InterRob Robotics Information Model – proposed extensions to STEP. Diagram 2 of 3

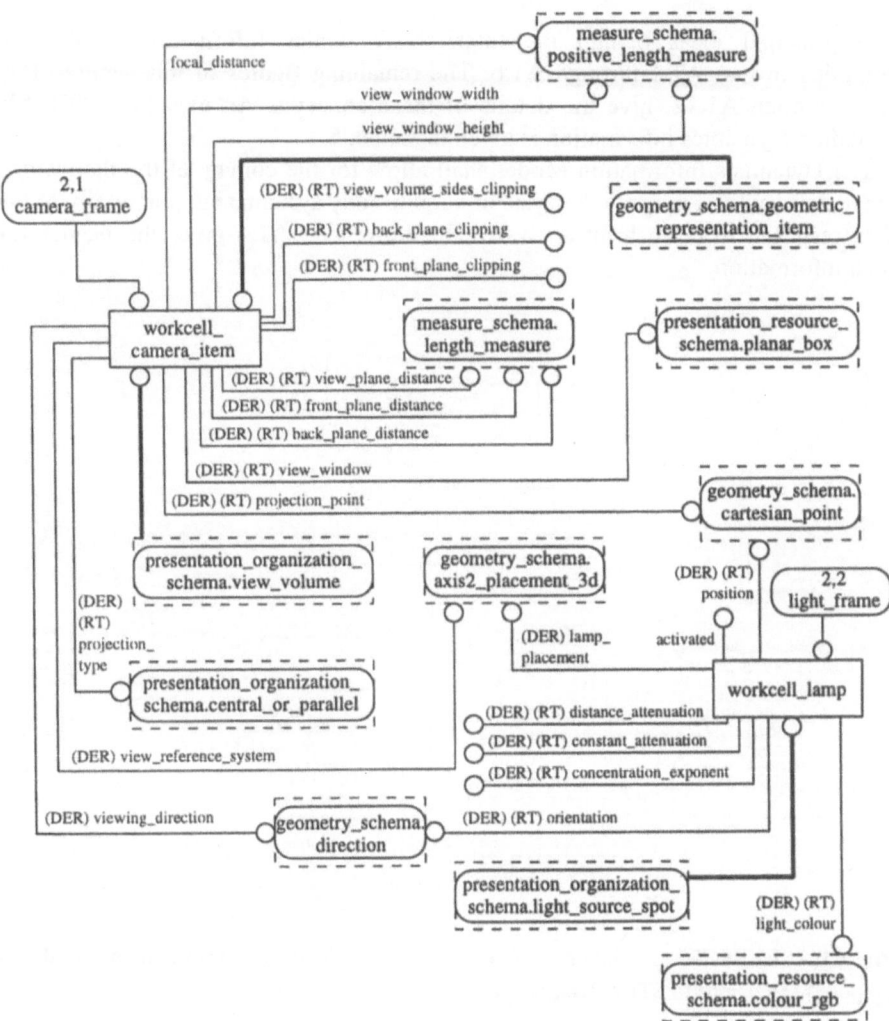

Fig. A1.3. EXPRESS-G diagram of the InterRob Robotics Information Model – proposed extensions to STEP. Diagram 3 of 3

The InterRob Dynamics Information Model

The principal ideas behind the InterRob Dynamics Information Model are compiled in Fig. A1.4 through A1.6. The remaining figures of this section, Fig. A1.7 through A1.12, give the details of these concepts. An example STEP file including Dynamics information is given in Annex 5.

The Dynamics Information Model shall allow for the convey of the data which are necessary for a dynamic analysis of a multi-body system, and, potentially of the data resulting from such an analysis. Therefore, Fig. A1.4 gives the header for such information.

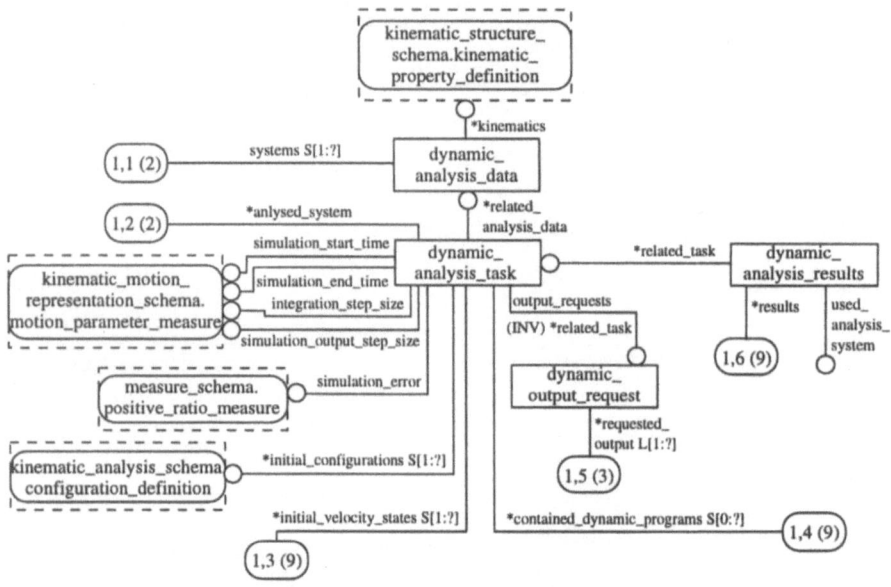

Fig. A1.4. EXPRESS-G diagram of the InterRob Dynamics Information Model – proposed extensions to STEP. Diagram 1 of 9

Figure A1.5 defines the constituents of a multi-body system seen from the dynamics point of view. There is at least an *inertia structure* containing the definitions of mass properties, and an *interact structure* by which the constraints and other interactions between the various bodies of the system are defined. Additionally, there may be any number of *subsystems* which, for the analysis under consideration, act like black boxes.

The top three quarters of Fig. A1.6 deal with the details of the *interact structure*. The interactions are treated as specialised transformations between the bodies involved. This concept includes kinematic constraints as an even more specialised case.

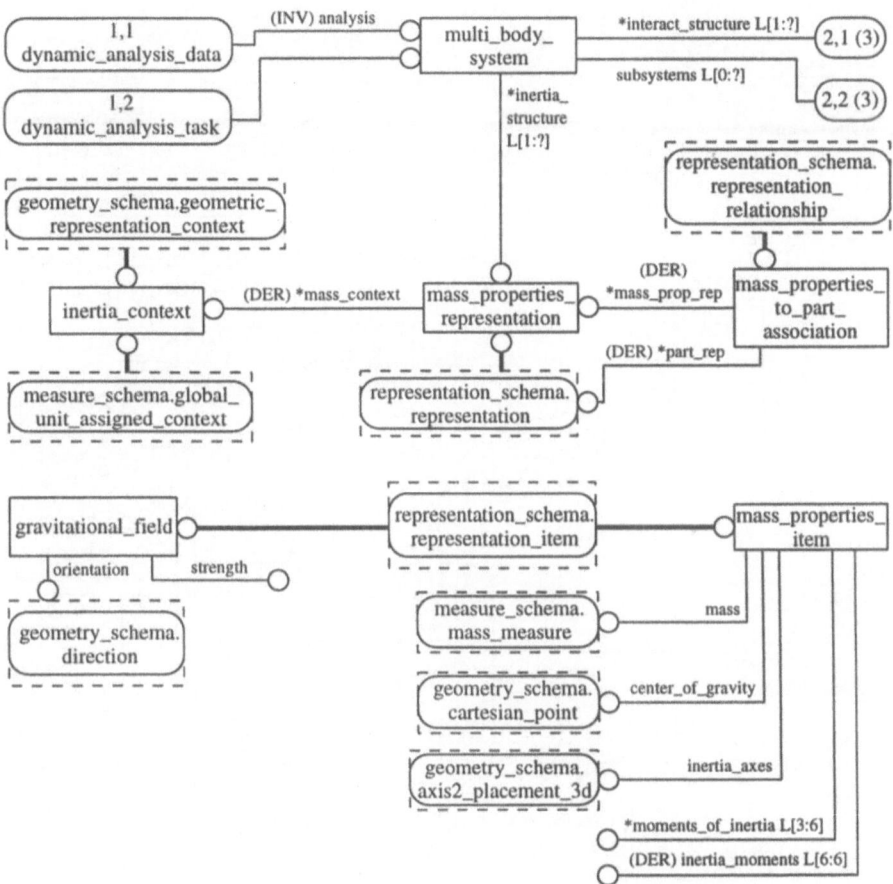

Fig. A1.5. EXPRESS-G diagram of the InterRob Dynamics Information Model – proposed extensions to STEP. Diagram 2 of 9

Besides the kinematic constraints, the Dynamics Information Model knows four kinds of *dynamic interactions*: forces, states, evaluators, and inputs. As the most obvious interactions, the *forces* do not need further explanation. *States* define constant values for an interaction, e.g., as the initial conditions. *Evaluators* ask for the interaction values as resulting from the analysis; they are required to be updated each time when the system internal state is output in the course of the analysis. *Inputs* are externally defined interaction values which may be time dependent or not; as seen from Fig. A1.8, the input values may address states or forces, and, hence, they may reference the corresponding elements of the *states* or *forces* interaction lists.

All the *interaction* constructs are embedded, as usual in STEP, in appropriate representations in order to provide them with suitable context definitions.

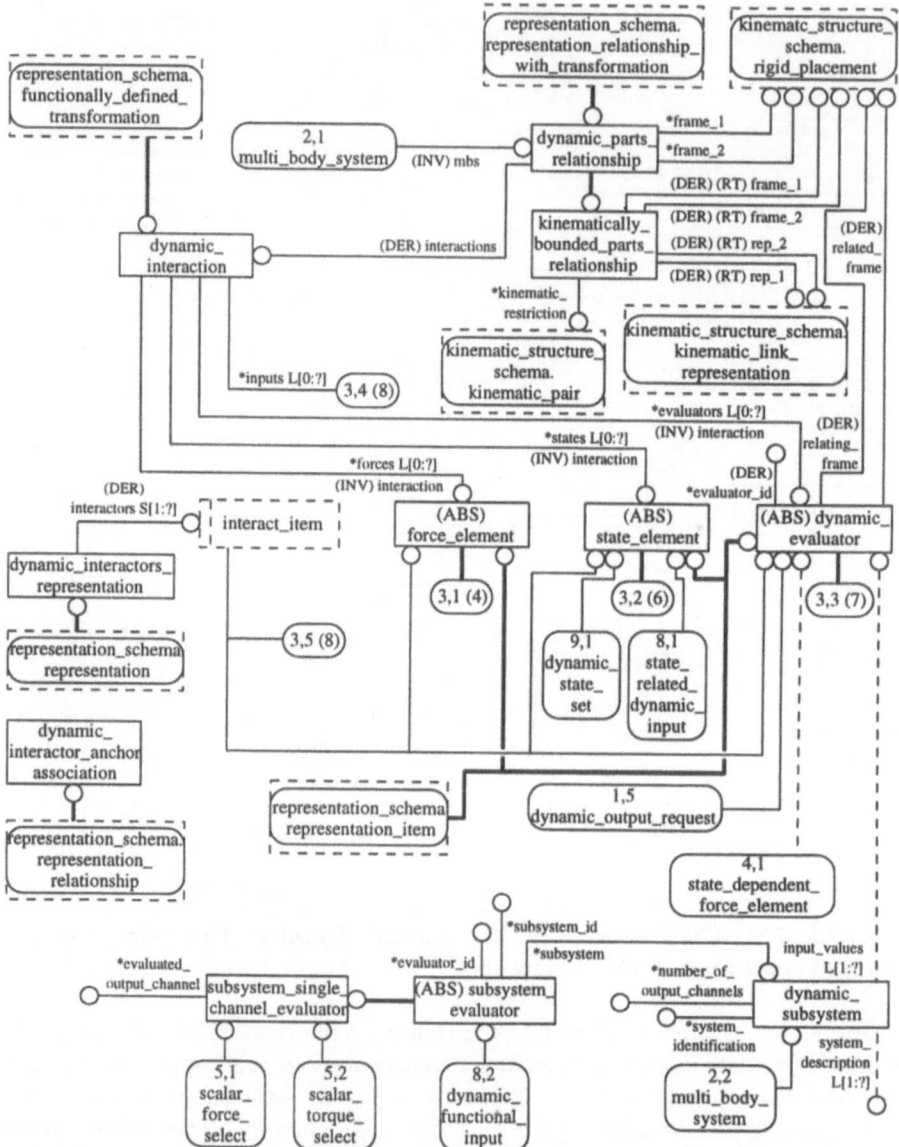

Fig. A1.6. EXPRESS-G diagram of the InterRob Dynamics Information Model – proposed extensions to STEP. Diagram 3 of 9

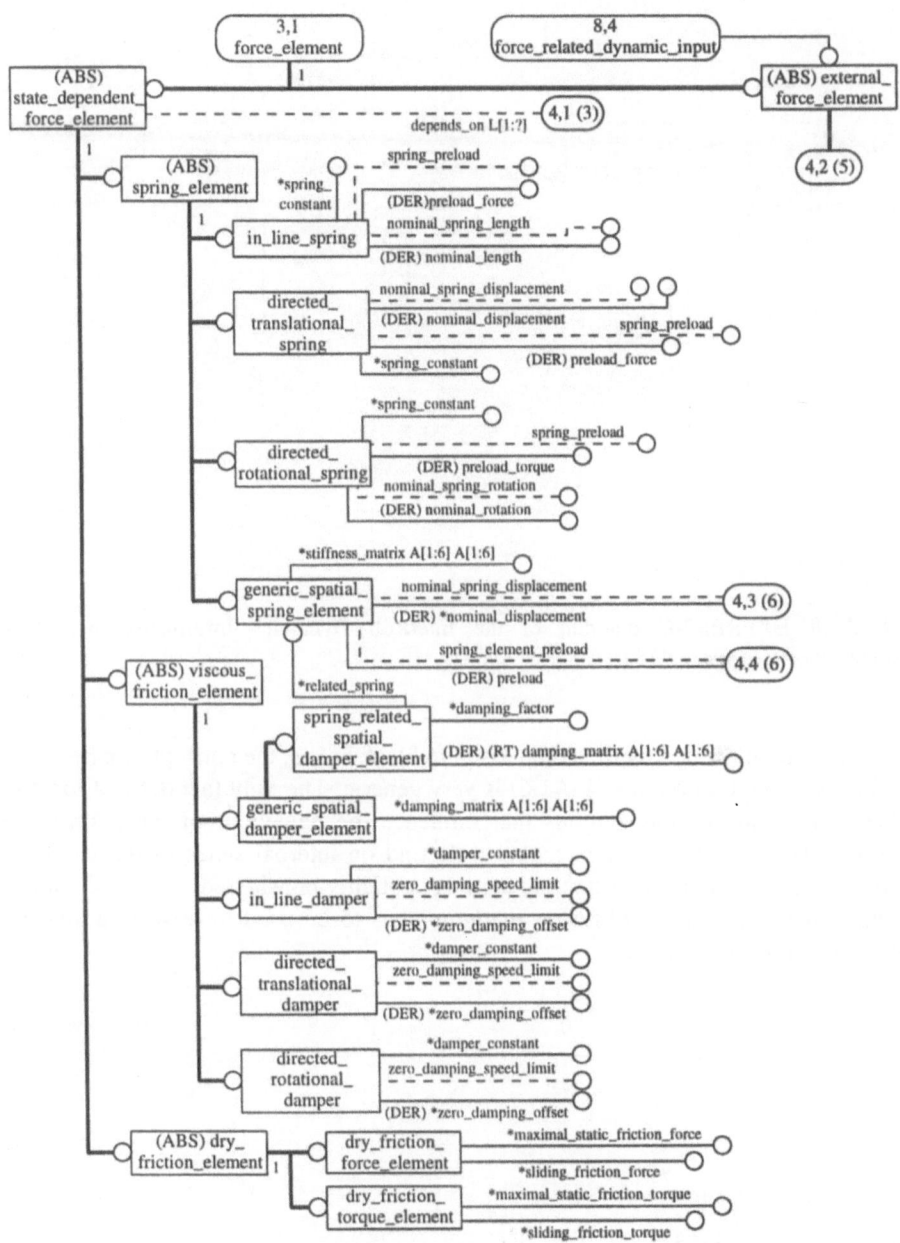

Fig. A1.7. EXPRESS-G diagram of the InterRob Dynamics Information Model – proposed extensions to STEP. Diagram 4 of 9

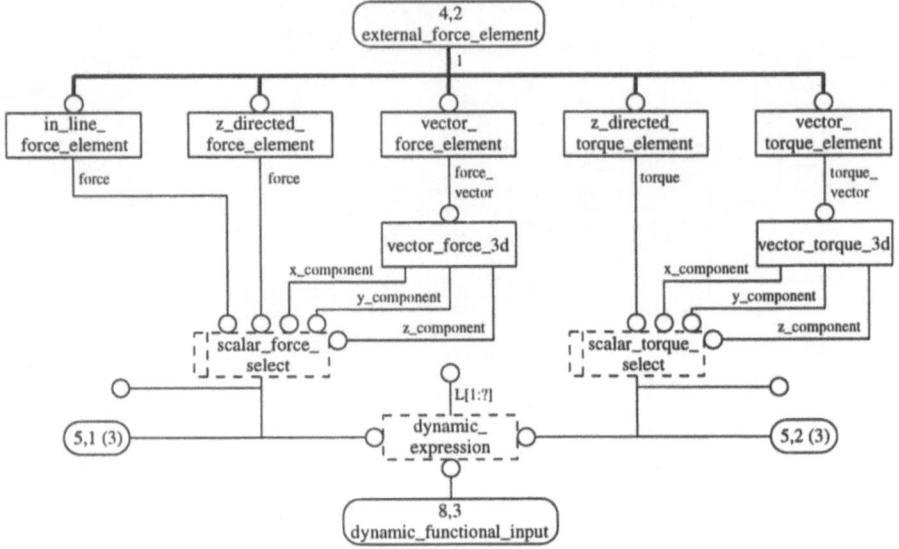

Fig. A1.8. EXPRESS-G diagram of the InterRob Dynamics Information Model – proposed extensions to STEP. Diagram 5 of 9

As already indicated by the comparison to black boxes, the concept of a *dynamic subsystem* (see bottom of Fig. A1.6) is very generic. The only fact defined for it is that it has one or more outputs that influence the behaviour of the multi-body system. These outputs may or may not depend on internal states of the analysed system. If, in turn, such an internal state is explicitly considered to depend on the output of a subsystem, *subsystem evaluator* has to be used to provide a kind of *dynamic functional input*.

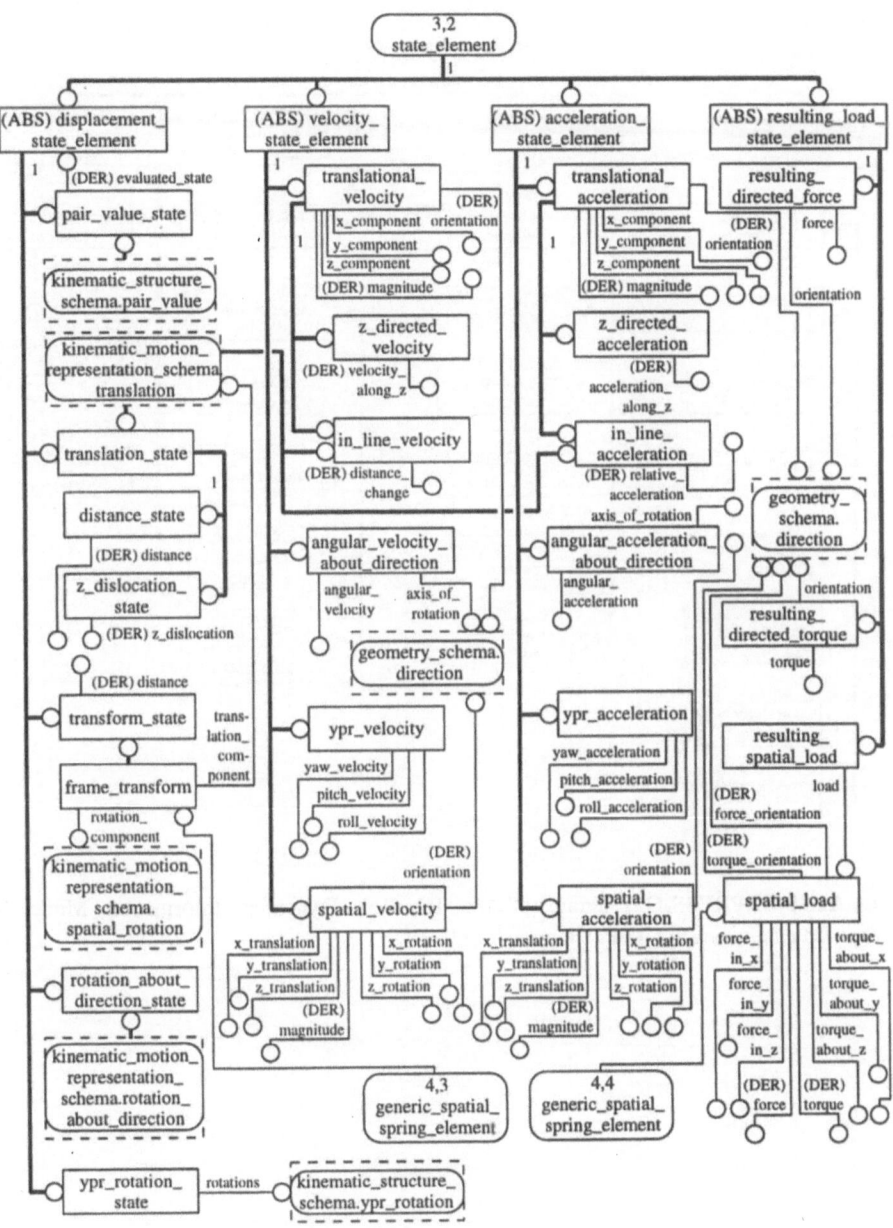

Fig. A1.9. EXPRESS-G diagram of the InterRob Dynamics Information Model – proposed extensions to STEP. Diagram 6 of 9

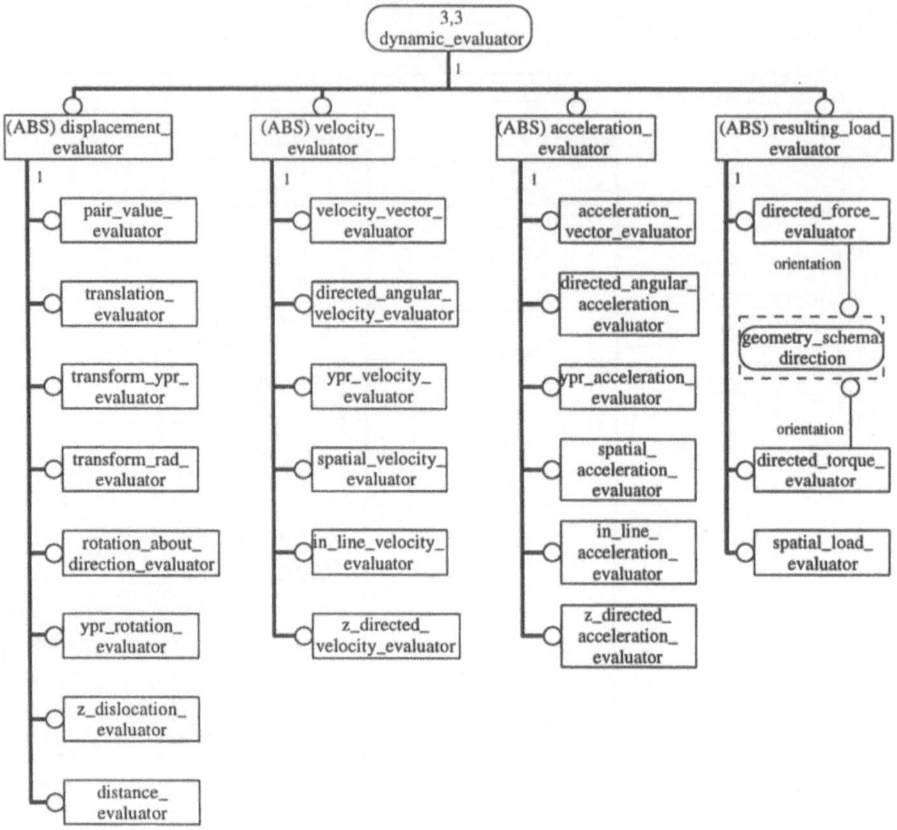

Fig. A1.10.EXPRESS-G diagram of the InterRob Dynamics Information Model – proposed extensions to STEP. Diagram 7 of 9

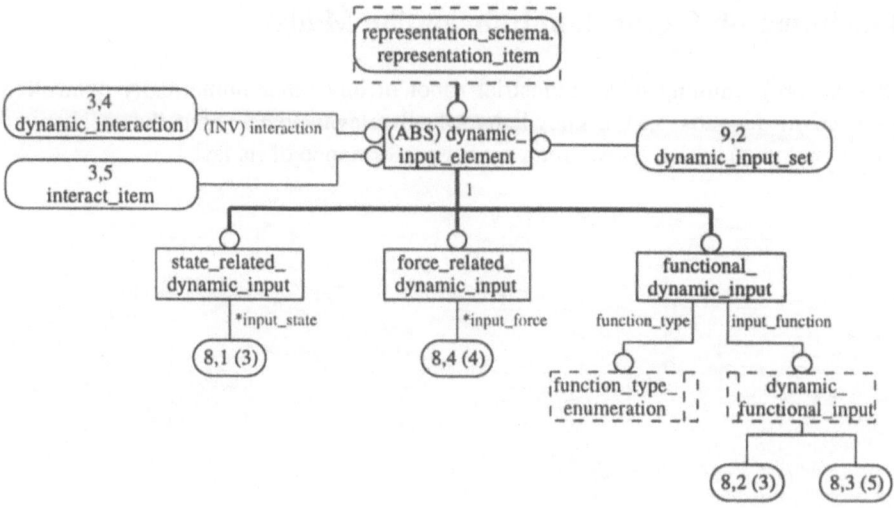

Fig. A1.11. EXPRESS-G diagram of the InterRob Dynamics Information Model – proposed extensions to STEP. Diagram 8 of 9

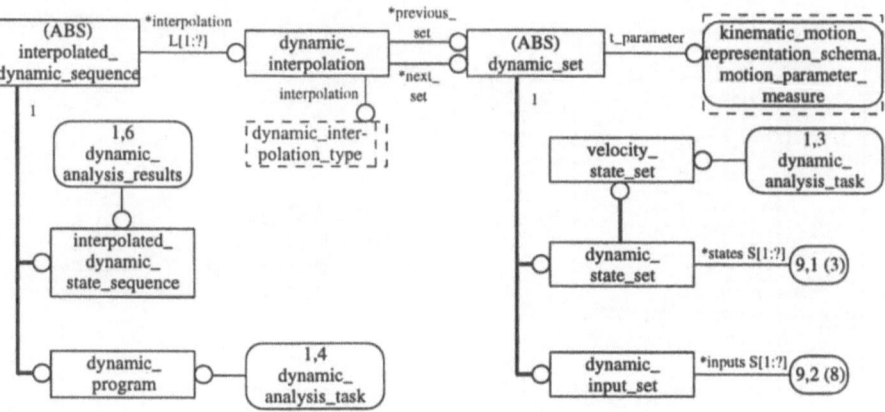

Fig. A1.12. EXPRESS-G diagram of the InterRob Dynamics Information Model – proposed extensions to STEP. Diagram 9 of 9

The InterRob Calibration Information Model

Off-line programming of an industrial robot or any other numerically controlled mechanism depends on the knowledge of all relevant data about the mechanism with a precision which allows for a proper performance of its task.

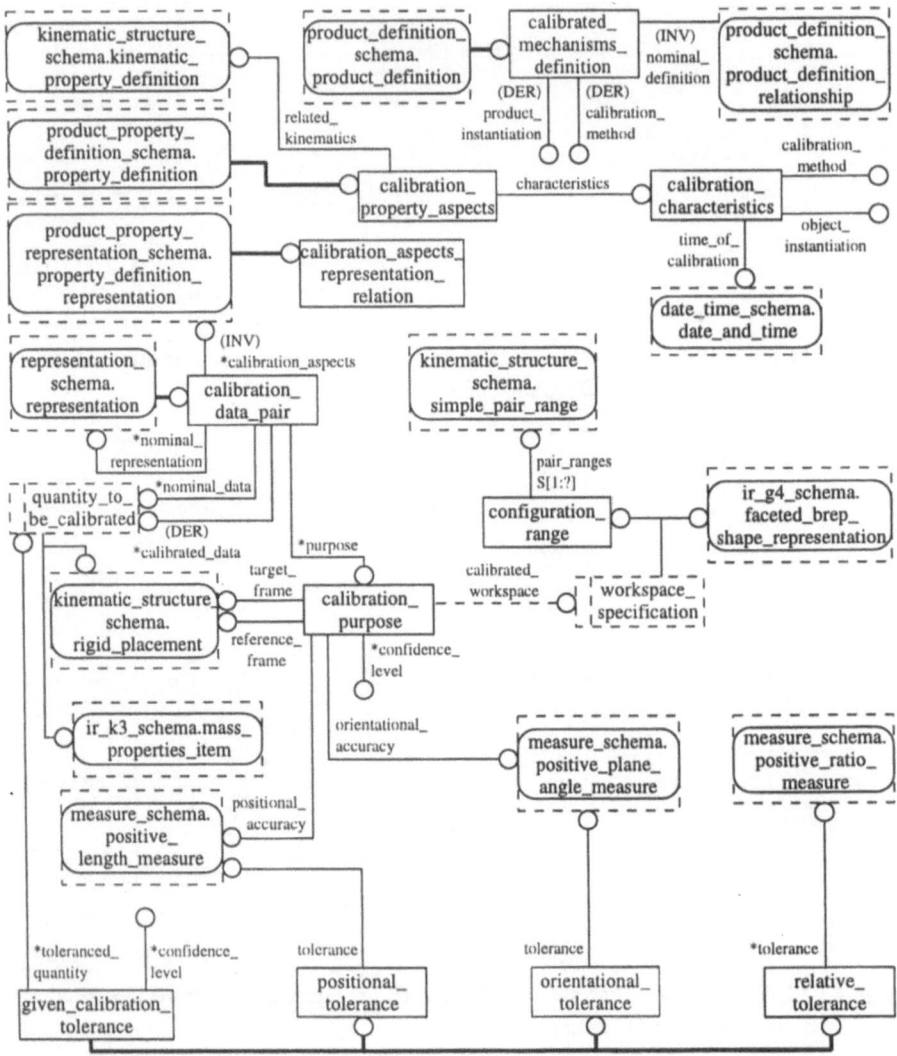

Fig. A1.13.EXPRESS-G diagram of the InterRob Calibration Information Model – proposed extensions to STEP. Diagram 1 of 1

However, due to various inaccuracies, real engines do not always meet their nominal data. Hence, it has become practice to calibrate the single specimens of a mechanism type, yielding a set of modified mechanism data which are more suitable to predict the mechanism data than the nominal data.

The InterRob Calibration Information Model (see Fig. A1.13) provides for calibrated data to replace nominal positioning information and nominal mass properties. This is a rather limited area of calibration data which have been choosen as prototypical examples, in view of the fact that up to now there is not yet a concept in STEP how to represent alternative data.

Again, the information model has links to the product structure data. It is this level where the calibration data are associated to a specific item of a given 'class' of mechanisms. The *calibration data pair* identifies the nominal data and the calibrated data which replace them. Furthermore, calibration data may be valid only for a portion of the workspace of the mechanism; therefore, the *calibration purpose* allows to define the spatial limits of applicability and the *target frame* with respect to which the calibration is to be or has been performed.

Finally, tolerance values allow to estimate the degree to which the calibration purpose is being met.

The InterRob Control Information Model

The generic motion controller model, identified and applied in InterRob, is shown along with its natural surroundings in Fig. A1.14.

In essence, a robot motion controller behaves much like a conventional controller; it seeks to minimise the error signal obtained by the comparison between the desired reference input data and the data that are fed back from measurements of the motions and forces, etc. realised by the controlled object. However, in contrast to most conventional controllers the motion controller consists, roughly speaking, of six fairly complicated functions to accomplish its task. These six functions are represented as functional modules of the motion controller in Fig. A1.15.

The six functions do the tasks needed to transform the actual input reference robot manipulator program from program statements into control signals for the servo motor drives that actuate the robot movements controlled by external and internal sensor feed-back.

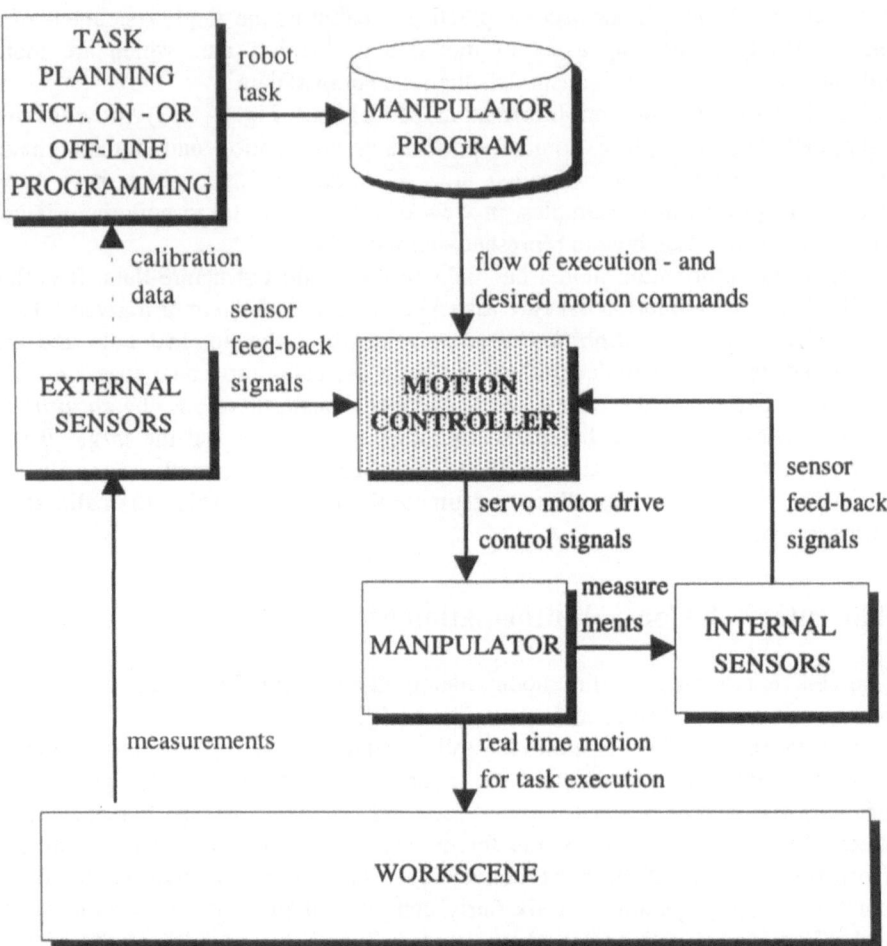

Fig. A1.14. The generic motion controller used in InterRob.

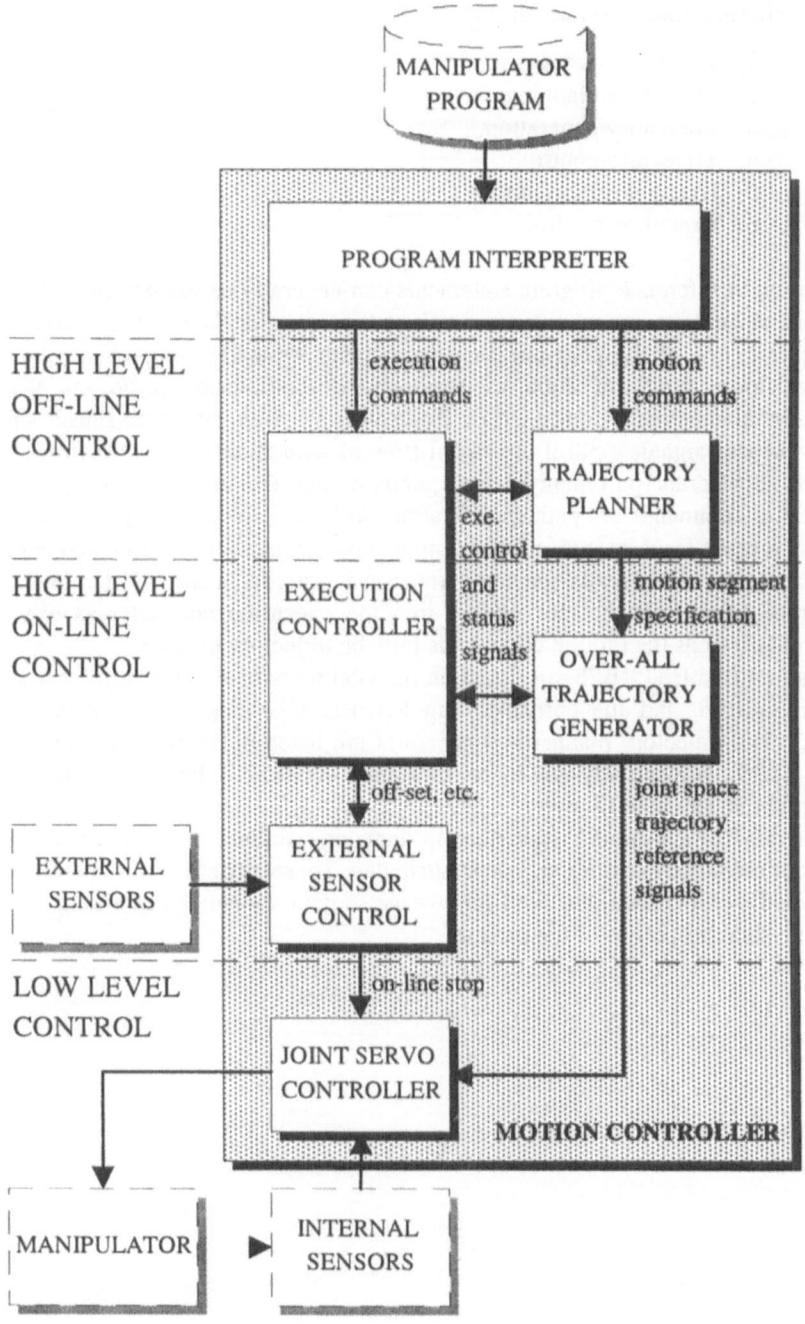

Fig. A1.15. The internal function modules of the generic motion controller.

The six functional modules are:

1. Program interpretation.
2. Trajectory planning.
3. Trajectory generation.
4. Trajectory control.
5. External sensor control.
6. Execution control.

The input reference program statements can generally be divided into execution commands and movement commands. Execution commands are used to control the execution flow of the program; these statements typically comprehend the well-known loop structures such as while-do, do-until, and for-do, as well as conditional statements controlled by Boolean expressions based on sensor signals. These sensor signals control the signal-flow of communication between the robot and its peripherals (part manipulators, gantry cranes, belt conveyors, etc.).

Motion commands are path specifications in Cartesian space or joint space such as move point-to-point, move linear, move circular, move uncoordinated, etc. The program interpreter reads the input reference program, converts it into internal program commands that are written into the execution controller as execution commands, or as the motion commands into the trajectory planner.

A simplified EXPRESS specification for vital parts of robot controller is shown in Fig. A1.16, and the corresponding EXPRESS-G diagram is shown in Fig. A1.17. The trajectory planner and generator are listed in EXPRESS in Fig. A1.18 and A1.19, respectively. A more complete EXPRESS listing of the generic controller is shown in [Soerensen 1995].

The EXPRESS model specification language defines the data structure of product models by entities and their attributes. These attributes of the entities, the so-called member data, are used for storage of data structures describing the state and physical properties of the product.

```
EXPRESS specification
*)
ENTITY generic_controller;
  contr_DOF : INTEGER;
  interpreter : program_interpreter;  planner :
trajectory_planner;
  generator : trajectory_generator;
  controller : ARRAY[1:contr_DOF] OF joint_controller;
external_sensor_controller : ARRAY [0:?] OF
virtual_sensor;  e_controller : execution_controller;
RRS_initialize : externally_defined_item;
  RRS_reset : externally_defined_item;
  RRS_terminate : externally_defined_item;
  RRS_get_robot_stamp : externally_defined_item;
  rcs_data : rrs_rcs_element;
END_ENTITY; (*
```

Fig. A1.16. Simplified EXPRESS specification of a robot controller

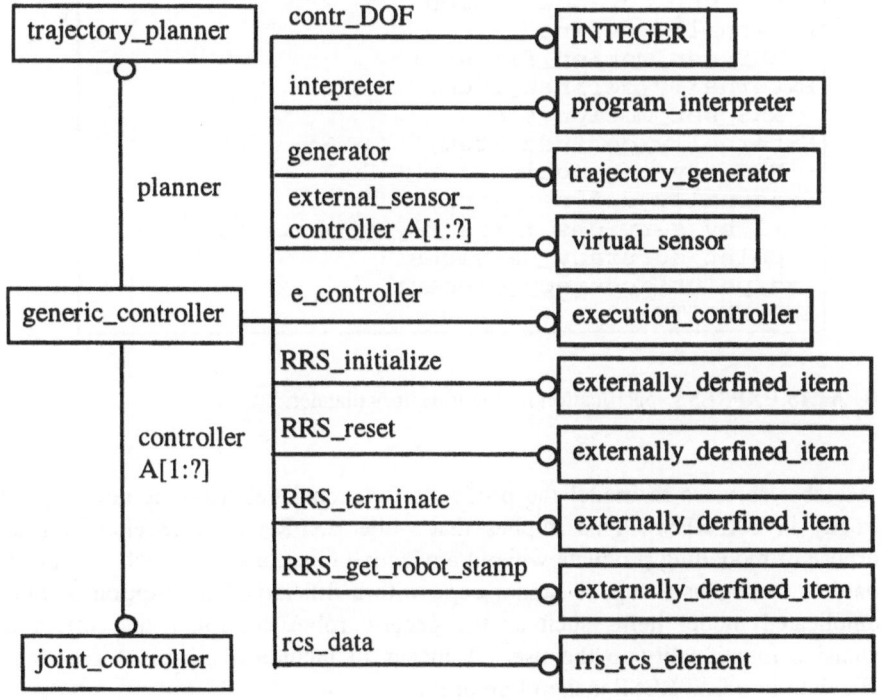

Fig. A1.17. Simplified EXPRESS-G diagram for the robot controller.

```
EXPRESS specification
*)
ENTITY trajectory_planner;
  joint_space          :
joint_space_trajectory_planner;
cartesian_space        :
cartesian_space_trajectory_planner;
kinetic_constraints : OPTIONAL
kinetic_computation_module;
RRS_set_initial_position :
Externally_defined_item;
  RRS_set_next_target :
Externally_defined_item;
  RRS_get_next_step :
Externally_defined_item;
  RRS_select_motion_type :
Externally_defined_item;
  RRS_select_target_type :
Externally_defined_item;
  RRS_select_trajectory_mode :
Externally_defined_item;
  RRS_set_advance_motion :
Externally_defined_item;
  RRS_set_motion_filter :
Externally_defined_item;
  RRS_set_motion_time :
Externally_defined_item;
  RRS_reverse_motion :
Externally_defined_item;
  flyby_services : rrs_flyby_element;
  point_accuracy_services :
rrs_point_accuracy_element;
END_ENTITY; (*
```

Fig. A1.18. EXPRESS specification for the trajectory planner.

Member services to model the product's functional behaviour is not included directly in EXPRESS. This implies that while EXPRESS is excellent for the purpose of modelling products without any active functionalities, such as a black-board, a carpet, or a cup, it cannot cope with a full model description of more complicated model items such as the generic robot controller that offers an extensive functionality to the user. A means to include a model of functional behaviour into EXPRESS is therefore needed.

```
EXPRESS specification:
*)
ENTITY externally_defined_item;
   item_source : source_item;
   source      : external_source;
END_ENTITY; (*
```

Attribute definitions:

item_source: is the definition of the referent source_item i.e. the name of the function defining the attribute service, e.g. 'get_inverse_kinematics'.
source: is an external_source (e.g. a file) which contains a source of product data that is not defined in the application protocol to which the exchange conforms, e.g. source = 'controller_services.cpp'.

```
EXPRESS specification:

*)
ENTITY external_source;
   source_id : source_item;
END_ENTITY;
(*
```

Attribute definitions:
source_id: the identification of the external source.

```
EXPRESS specification:
*)
TYPE source_item = SELECT (identifier, message);
END_TYPE; (*
```

Attribute definitions:
message: a communication which is addressed to a system to trigger some action. The result of such an action is an externally_defined_item.
identifier: An alphanumeric string which allows an individual thing to be identified.

```
EXPRESS specification:
*)
TYPE identifier = STRING;
END_TYPE;TYPE message = STRING;
END_TYPE;
(*
```

Fig. A1.19. The EXPRESS specification of the externally_defined_item entity, including the entities and type definitions used by the externally_defined_item entity to provide the interface specification with a mechanism to include a description of the products functionality.

In order to describe the functional behaviour of product models in EXPRESS and other needs as well, a so-called *externally_defined_item* has been introduced [ISO 10303-41, 1994], see Fig. A1.19.

The externally_defined_item is used in the InterRob specification to incorporate the member services, to extend the classic EXPRESS definitions to also include a functional description of the entities that defines the robot controller. This way, the entity definition becomes very close to the definition used for objects in object-oriented languages as shown for the generic_controller definition in C++ in Fig. A1.20. The approach therefore not only adds the description of the product's functionality into the specification, but it also provides the benefit of a smoother integration between the specification and schema-driven processors based on an object oriented computer programming language such as C++, Eiffel, and SmallTalk.

Most of the services defined for the generic robot controller are written to support the so-called RRS[1] -interface specification [Bernhardt 1994]. The RRS interface is a result of a corporate project with participation from several important manufacturers[2], primarily from Europe.

The main purpose of the RRS interface is to enable the use of the same original software functions in simulation that are applied in the real robot controller to increase the fidelity of the simulation results.

The RRS interface is defined by prototype declarations of 57 functions for motion planning and inverse/direct kinematics. In the InterRob specification the RRS functions start with an 'RRS_' prefix to ease the identification of the RRS functional interface within the controller product model.

This Annex has given a short survey of the generic STEP based motion controller model that has been made to allow free STEP based storage, and exchange of controller model between CACSD systems and CAR systems without information losses that can cause a significant model degeneration. Please refer to references [Bernhardt 1994], [EXPRESS 1992], [STEP41 1994], and [Sørensen 1995] for a more comprehensive and detailed description on the topics addressed in this Annex.

---

[1] RRS is acronym for Realistic Robot Simulation.

[2] 10 car-, 6 robot-, 1 aeroplane-, and 2 robot simulation systems vendors, and one research organisation.

```
class generic_controller {
  const int contr_DOF;
public:
  program_interpreter*    interpreter;
  execution_controller*   e_controller;
  trajectory_planner*     planner;
  trajectory_generator*   generator;
  joint_controller*       control[contr_DOF];
  virtual_sensor**        sensor_control;

  void RRS_initialize(int RobotNumber,
                      PathName RobotPathName,
                      PathName ModulePathName,
                      String ManipulatorType,
                      int CARRRSVersion,
                      int Debug, int* Status,
                      bitstring2* RCSHandle,
                      int* RCSRRSVersion,
                      int* RCSVersion,
                      int* NumberOfMessages);

  void RRS_reset(bitstring2 RCSHandle, int ResetLevel,
                 int* Status,   int* NumberOfMessages);

  void RRS_terminate(bitstring2 RCSHandle,
                     int* Status);

  void RRS_get_robot_stamp(bitstring2 RCSHandle,
                           int* Status,
                           string* Manipulator,
                           string* Controller,
                           string* Software);

  rrs_rcs_element rcs_data();
};
```

Fig. A1.20. Class definition in the C++ programming language of the generic_controller. (See Fig. A1.16 and A1.17 for the EXPRESS and EXPRESS-G specification of the generic_controller).

Annex 2   STEP Processors Developed in InterRob

T. Sørensen et al.
Danmarks Tekniske Universitet, Instituttet for Styreteknik
Bygning 424, DK-2800 Lyngby, Denmark

This Annex describes the processors developed in the InterRob project. Processor developments are based upon so-called schemas that define models of the information the processors have to transfer. Several STEP schemas have been developed in InterRob and the primary goals of the processors are to test and verify the consistency of these schemas. The schemas have been denoted by the identifiers S1, G3, G4, and K1 to K5 respectively, and the capabilities of the processors are indicated by the identifiers of the schema(s) it supports:

S1 indicates that the processor supports the *Product structure, template, and instance* schema.
G3 indicates that the processor supports the *Face-based surface model* schema.
G4 indicates that the processor supports the *Faceted B-rep.* schema.
K1 indicates that the processor supports the *Basic kinematics* schema.
K2 indicates that the processor supports the *Robotics model* schema.
K3 indicates that the processor supports the *Manipulator dynamics* schema.
K4 indicates that the processor supports the *Robot controller model* schema.
K5 indicates that the processor supports the *Robot calibration* schema.

The processors have been written for seven different CAD- and robot off-line programming systems and are listed in the two tables shown below. The systems involved are:

ADAMS, a software package for simulating the force and motion behaviour of mechanical systems, developed and sold by Mechanical Dynamics, Inc., Ann Arbor, Michigan, and widely used in various mechanical industries.

Bravo3, a mechanical 3D CAD system using the CSG approach for its solid models. It is sold by *Applicon*. ROBOT is an enhancement to Bravo3 developed by FZK in Germany. ROBOT allows to define kinematics and - to some extent - robotics models in the Bravo3 environment. Thus, the shape models are genuine Bravo3 models, whereas the kinematics and robotics information is stored in ROBOT models. ROBOT models can be visualised and "simulated" interactively on the screen.

CATIA, a general purpose Computer Aided Design and Manufacturing (CAD/CAM) system developed by Dassault in France and marketed world-wide by IBM. Besides the basic modules for user interaction, database management and graphics, together with a 2D and 3D wire-frame geometry modeller, the system features a series of application modules like DRAFTING, SURFACES, SOLIDS, KINEMATICS, ROBOTICS, etc.

GRASP, a full 3-D software package designed to model robots, other kinematic structures, and their surrounding environment. It has its own solid modeller and build-in facilities for solving inverse kinematic equations for a large range of kinematics structures found in most industrial robots. Its use is widespread as a robot cell planning and visualisation system, and a management tool to help determine the optimum solution for a particular automated task.

KISMET, a simulation system for kinematic and dynamic mechanisms developed at FZK in Germany. Though there are some modelling capabilities inside KISMET, the models are usually imported from a CAD system. In some cases, only the shape models may be imported, whereas the kinematics and dynamics properties may be added or changed by the KISMET user.

ROPSIM, a true CIME sub-system for model driven real-time simulation of the dynamic behaviour of mechanical systems, in particular robots, for given loads and control laws.. ROPSIM is written in C++ and developed at Control Engineering Institute, DTU, Denmark. ROPSIM stands for off-line RObot Programming and SIMulation system. More details of ROPSIM are given in Annex 3.

SISL, a Spline Library based on the NURBS format (Non Uniform Rational B-Splines) developed by SINTEF, Norway. In InterRob SISL is used within a geometry conversion system developed at SINTEF.

STEP pre-processors developed in InterRob

SYSTEM	SUPPORTED SCHEMAS						
ADAMS			K1		K3		
BRAVO3	G3	G4	K1	K2			K5
CATIA		G4	K1	K2	K3		
GRASP	G3	G4	K1	K2			K5
KISMET	G3	G4	K1	K2			K5
ROPSIM					K3	K4	
SISL	G3						

STEP post-processors developed in InterRob

SYSTEM	SUPPORTED SCHEMAS						
ADAMS			K1		K3		
GRASP	G3	G4	K1	K2			K5
KISMET	G3	G4	K1	K2			K5
ROPSIM		G4	K1		K3	K4	
SISL	G3						

Annex 3   Simulation of Robot Kinetics and Control

Thomas Horsch[1], Uri Kroszynski[2]
[1] Reis Robotics
P.O. Box 11 01 61, D-63777 Obernburg, Germany
[2] Danmarks Tekniske Universitet, Instituttet for Styreteknik
Bygning 424, DK-2800 Lyngby, Denmark

This Annex gives a short overview about further work developed in Work Package 2 of InterRob regarding the simulation of robot dynamics and control. This work was performed with the software package ROPSIM (**Ro**bot **O**ff-line **P**rogramming and **SIM**ulation system). This system is intended for analysis and simulation of the dynamic behaviour of mechanical systems, in particular robots, for given loads and control laws. The ROPSIM system was developed between 1990 and 1992 at the Control Engineering Institute (IfS) of the Technical University of Denmark (DTU).

In InterRob ROPSIM was used as a platform for STEP model data exchange, not only for geometry and kinematics, but also in the areas of dynamics and control. In fact no other software was available that covered the control dynamics features needed in the project. It was decided to undertake a simulation of the Reis RV6 robot. The relevant data needed were kindly provided by the robot manufacturer Reis.

This Annex presents the methods used in ROPSIM for the dynamic modelling and simulation of the Reis RV6 robot. The simulation task was carried out at the Control Engineering Institute, Technical University of Denmark and is documented as a confidential report in [Nielsen, 1994].

A robot system in ROPSIM consists of three components:

- the manipulator system, namely models of the articulated mechanical assembly of the arms and tool (links) and the motors and gears at the articulations (joints).
- the controller system, and
- the robot program, i.e. the specification of the task to be performed.

Manipulator model

For the manipulator model, the descriptions of the geometric shape and kinematic structure of the robot are imported into ROPSIM from a generic CAD system employing a STEP neutral file. The mass properties of each arm (link) of the manipulator are modelled as concentrated parameters. These include the mass, the location of the centre-of-mass, the inertia axes, and the moments of inertia. These data are provided by the robot manufacturer, and added manually in the STEP file according to a specified entity format, with reference to the corresponding link. Similarly, entities for motor and gear are added to the STEP file for each actuated joint articulation. Further information like the tool and load mass properties as well as the orientation of the manipulator with respect to the ground in order to incorporate gravity impacts are added to the STEP file.

ROPSIM regards the manipulator as a system where the links are coupled, thereby influencing each other. The manipulator dynamics are governed by a non-linear matrix differential equation relating the torque applied at each joint with the angle, angular velocity, and angular acceleration variables at that joint (and at the other joints as well). The coefficients in the matrix equation represent the inertia, Coriolis, and gravity dynamics, and depend on the values of the variables at the joints in the system.

The manipulator model can be considered at three levels of detail: fully coupled, weakly coupled, and uncoupled (see [Horsch 1995] for further details).

Controller model

The controller model includes Cartesian trajectory planning, inverse kinematic transformator, joint trajectory planning, and the servo (joint) models. A generic motion controller model, which is assumed in ROPSIM consists of four components:

- the *Cartesian trajectory planner* includes the Cartesian path planner and interpolator
- the *inverse kinematic transformator* that converts Cartesian co-ordinates of a selected point/frame, e.g., the TCP, to robot joint space
- the *joint trajectory planner* includes the joint path planner and interpolator
- the joint controller consists of the servo loop controllers

There are two control loops per servo axis in the Reis RV6 joint controller.

- the velocity controller which is modelled as a proportional integral (PI) controller
- the position controller which is modelled as a proportional (P) controller.

Robot program

The robot program is considered as a model of the robot task. ROPSIM supports primarily the ICR neutral description. Nevertheless, IRL programs can be integrated to ROPSIM, by means of an IRL to ICR translator, to drive the simulation (see also Annex 4). A corresponding translation of the program from IRL to the Reis RRL native language is needed in order to run the same task on the physical RV6 robot.

Results

Simulations results, which are based on robot programs expressed in IRL, were produced for the following steps:

- in the parameter calibration phase the transient responses due to step input applied on the servos
- simulations of the Cartesian path tracking error for a nominal square trajectory
- simulations of the Cartesian path tracking error for a nominal Schmid curve (popular robot test curve).

The actual path obtained on a physical robot set-up similar to the simulated one for the Schmid curve case showed good qualitative agreement between measured and simulated results. These results and comparisons are fully documented in [Nielsen 1994].

Step responses of joint servos

The third joint axis of the RV6 was chosen for the step response simulation runs. This is because there are other joint axes both before and after, so the coupling influence from the other axes is expected to be representative. The transient responses are simulated with all other joint references set to zero. The step variation in the third axis position reference was set to 0.001 radians in order to avoid saturation in the AC-motor thereby ensuring the linearity of the system. Transient responses of the joint states (angular position, velocity, and acceleration) were computed and plotted. When using the uncoupled approximation, only joint three exhibits a response and all other joint states remain zero. Of particular interest are the plots of the joint states due to a step input to the joint servo on the third axis of the RV6 robot using the weakly coupled approximation. The transient responses from both ROPSIM and MATLAB were illustrated in [Nielsen 1994] showing very good agreement.

Square path and Schmid curve tracking error of the tool centre point (TCP)

These tasks were designed to show the relevance of dynamic effects. The results obtained from the ROPSIM simulation show the Cartesian tracking errors of the tool-centre-point (TCP), the states of the manipulator joints (position, velocity, acceleration), and the states of the controller (position and velocity error of the joints and TCP reference). The following simulation results have been carried out:

- TCP path tracking errors
- angular position, velocity, and acceleration at each individual joint
- position and velocity reference to the individual joint servo controller
- position and velocity errors in the individual joint servo controllers, and
- Cartesian TCP reference relative to the robot base.

The simulation results indicate that dynamic effects become more relevant at higher velocities. Sudden changes in the path direction, e.g. sharp corners, are tracked showing the deviations from the nominal path and overshoot. A full description of results is found in [Nielsen 1994]. These results have been compared with the plots of a physical robot executing the Schmid path. The qualitative agreement was remarkable and finally proved the effectiveness of ROPSIM.

Annex 4 IRL Processors developed in InterRob

P. Sorenti
BYG Systems Ltd, William Lee Building, Highfields Science Pard
Nottingham NG7 2RQ, United Kingdom

This Annex presents a list of IRL processors developed in InterRob. The processors are based on the German standard described in DIN 66312 and have been implemented on several platforms. For further information including development tools used to implement the processors, the way how to invoke them and their limitations refer to [Horsch 1994] (IRL prototype specification of InterRob).

- IRL → RRL
  Converts IRL programs to the **Reis Robot Language (RRL)**
- RRL → IRL
  Converts RRL programs to IRL
- IRL → ARLA
  Converts IRL programs to the ASEA robot programming language ARLA version 3.
- ARLA → IRL
  Converts ARLA programs to IRL.
- GRASP → IRL
  converts robot programs stored in the internal data base of GRASP release 8.6.5 to IRL.
- IRL → GRASP
  Converts a limited set of statements of IRL to GRASP.
- IRL → ICR
  Converts IRL programs to ICR (**I**ntermediate **C**ode for **R**obots) for the use in ROPSIM (**R**obot **O**ff-line **P**rogramming and **SIM**ulation system).

Annex 5 Examples of STEP Exchange Files in InterRob

F. Høgberg et al.
SINTEF Informatics
P.O. Box 124 Blindern, N-0314 Oslo, Norway

Example STEP File Including Geometry and Kinematics

This file has been shortened for sake of clarity. Intermediate instances in sequences of similar instances have been omitted, as indicated by ellipses.

```
ISO-10303-21;
HEADER;
FILE_DESCRIPTION( ( 'pencil_robot' ), '1' );
FILE_NAME( '/usr1/people/lutz/src/newenv.stp', '1995-10-26
T11:22:07',
     ('Richard Lutz' ),
     ('Forschungszentrum Karlsruhe GmbH Technik und Umwelt',
     'HIT',
     'Postfach 3640',
     'D-76021 Karlsruhe',
     'Phone : +49 7247 82-3128',
     'FAX   : +49 7247 82-2588',
     'e-mail: lutz@hit.fzk.de' ),
     '/htlsg6/people/lutz/src/robstt, RobDB 4.00',
     'Bravo/KISMET',
     'FZK/HIT' );
FILE_SCHEMA( ( 'IR_APPLICATION_SCHEMA' ) );
ENDSEC;
DATA;
/*    */
/* *** Global Units *** */
#100 = (LNGUNT( )NMDUNT(* )SUNT($,.METRE. ) );
#101 = (NMDUNT(* )PLANUN( )SUNT($,.RADIAN. ) );
/*    */
/* FACETED_BREP:    */
/* newenv_3 */
/*    */
#102 = &SCOPE    /* FCTBR */
  #103 = CRTPNT('point_103',(0.000000,0.000000,0.000000 ) );
  . . .
  . . .
  #110 = CRTPNT('point_110',(1.000000,0.100000,0.000000 ) );
  #111 = PLYLP('loop_111',(#103,#109,#108,#106 ) );
  #112 = FCOTBN('outer_bnd_112',#111,.T. );
  . . .
  . . .
```

```
   #121 = PLYLP('loop_121',(#104,#107,#110,#105 ) );
   #122 = FCOTBN('outer_bnd_122',#121,.T. );
   #123 = DRCTN('dir_123',(0.000000,-1.000000,0.000000 ) );
   #124 = DRCTN('dir_124',(0.000000,0.000000,-1.000000 ) );
   #125 = A2PL3D('axis2pl_125',#103,#123,#124 );
   #126 = PLANE('plane_126',#125 );
   #127 = FCSRF('face_srf_127',(#112 ),#126,.T. );
   ...
   ...
   #148 = DRCTN('dir_148',(0.000000,1.000000,0.000000 ) );
   #149 = DRCTN('dir_149',(0.000000,0.000000,1.000000 ) );
   #150 = A2PL3D('axis2pl_150',#104,#148,#149 );
   #151 = PLANE('plane_151',#150 );
   #152 = FCSRF('face_srf_152',(#122 ),#151,.T. );
   #153 = CLSSHL('shell_153',(#127,#132,#137,#142,#147,
       #152 ));
ENDSCOPE
FCTBR('B-rep_102',#153 );

#154 = &SCOPE    /* REPRESENTATION_MAP */
   #155 = &SCOPE    /* A2PL3D */
    #156 = CRTPNT('point_156',(0.00000,0.00000,0.00000 ) );
    #157 = DRCTN('dir_157',(1.000000,0.000000,0.000000 ) );
    /* Y = DRCTN('dir1581',(0.000000,1.000000,0.000000)); */
    #158 = DRCTN('dir_158',(0.000000,0.000000,1.000000 ) );
   ENDSCOPE
   A2PL3D('matrix_155',#156,#158,#157 );
   #159 = (GMRPCN(3 )GUAC((#100,#101 ) )RPRCNT('LCS',
       'Context for template 160' ) );
   #160 = FBSR('fbsr_template_160',(#102,#155 ),#159 );
ENDSCOPE
RPRMP(#155,#160 );
/*    */
#161 = &SCOPE    /* MAPPED_ITEM */
   #162 = DRCTN('dir_162',(1.000000,0.000000,0.000000 ) );
   #163 = DRCTN('dir_163',(0.000000,0.000000,1.000000 ) );
   #164 = DRCTN('dir_164',(0.000000,-1.000000,0.000000 ) );
   #165 = CRTPNT('point_165',(0.000000,0.000000,0.000000 ) );
   #166 =
CTO3('rep_item_166','cto3_166','trafo',#162,#163,#165,1.000,
       #164 );
ENDSCOPE
MPPITM('MPD_ITEM_411',#154,#166 );
/*    */
/* FACETED_BREP:   */
/* newenv_57 */
/*   */
#167 = &SCOPE    /* FACETED_BREP */
   ...
   ...
   #349 =
CLSSHL('shell_349',(#243,#248,.....,#333,#338,#343,#348 ) );
ENDSCOPE
FCTBR('B-rep_167',#349 );

#350 = &SCOPE    /* RPRMP */
```

```
   ...
   ...
ENDSCOPE
RPRMP(#351,#356 );
/*    */
#357 = &SCOPE    /* MPPITM */
   ...
   ...
ENDSCOPE
MPPITM('MPD_ITEM_417',#350,#362 );
/*    */
/* FACETED_BREP:   */
/* newenv_230 */
/*    */
#363 = &SCOPE    /* FCTBR */
   ...
   ...
   #550 =
CLSSHL('shell_550',(#434,#439,...,#534,#539,#544,#549 ) );
ENDSCOPE
FCTBR('B-rep_363',#550 );

#551 = &SCOPE    /* RPRMP */
   ...
   ...
ENDSCOPE
RPRMP(#552,#557 );
/*    */
#558 = &SCOPE    /* MPPITM */
   ...
   ...
ENDSCOPE
MPPITM('MPD_ITEM_423',#551,#563 );
/*    */
#564 = &SCOPE    /* RPRMP */
   ...
   ...
ENDSCOPE
RPRMP(#565,#570 );
/*    */
#571 = &SCOPE    /* MPPITM */
   ...
   ...
ENDSCOPE
MPPITM('MPD_ITEM_429',#564,#576 );
/*    */
/*    */
/* *** PRODUCT INFORMATION *** */
#577 = APPCNT(
    'InterRob model transfer / RobDB V4.0 / S1-G4-K1' );
#578 = APPRDF('DRAFT_PROPOSAL',
    'IR_APPLICATION_SCHEMA',1995,#577 );
#579 = PRDCNT('InterRob_model',#577,'Mechanical' );
#580 = PRDCT('newenv','ER','root_of_the_model',(#579 ) );
#581 = PRDFFR('VERS01','first_version',#580 );
#582 = PRPC('ROBOT','Modelled for InterRob',(#580 ) );
```

```
/*    */
/* *** workcell definitions for the kinematic model *** */
#583 = PRDFCN('Kinematic Model',#577,'Undefined' );
#584 = PRDDFN('Workcell_584','Kinematic model and
ground',#581,#583 );
/*    */
#585 = PRDFSH('Shape_585','LINK_Shape',#584 );
#586 = CHROBJ('environment_586','The ground of the kinematic
model' );
/*    */
/* *** The Kinematic Property Definition *** */
#587 = &SCOPE    /* KNPRDF */
  /*    */
  /* *** The Kinematic Ground Definition *** */
  #588 = &SCOPE    /* KNGRRP */
    #589 = (GMRPCN(3 )GUAC((#100,#101 ) )RPRCNT('WCS',
        'Context for the shape of the ground (#588)' ) );
    #590 = &SCOPE    /* CTO3 */
      #591 = DRCTN('dir_591',(1.00000,0.00000,0.00000 ) );
      #592 = DRCTN('dir_592',(0.00000,1.00000,0.00000 ) );
      #593 = DRCTN('dir_593',(0.00000,0.00000,1.00000 ) );
      #594 = CRTPNT('point_594',(0.00000,0.00000,0.00000 ));
    ENDSCOPE
    CTO3('rep_item_590','cto3_590','trafo',
        #591,#592,#594,1.000000,#593 );
  ENDSCOPE    / #590 /
  KNGRRP('kin_ground_rep_588',(#590 ),#589 );
  /*    */
  /* *** The Definition of Mechanism  newenv_M441 *** */
  #595 = &SCOPE    /* MCHNSM */
    /* Kinematic Link  KIN_LINK_442 : */
    #596 = KNMLNK( );
    /* Kinematic Link  KIN_LINK_443 : */
    #597 = KNMLNK( );
    /* Kinematic Link  KIN_LINK_444 : */
    #598 = KNMLNK( );
    /* Kinematic Link  KIN_LINK_445 : */
    #599 = KNMLNK( );
    /*    */
    /* Kinematic Joint  KIN_JOINT_446 : */
    #600 = KNMJNT(#596,#597 );
    /* placement of pair relative to first  link of joint */
    #601 = &SCOPE    /* A2PL3D */
      #602 = CRTPNT('point_602',(0.00000,0.00025,0.00010 ));
      #603 = DRCTN('dir_603',(0.000000,-1.000000,0.00000 ));
      /* Y = DRCTN('dir6041',(0.000000,0.00000,1.00000)); */
      #604 = DRCTN('dir_604',(-1.00000,0.00000,0.00000 ));
    ENDSCOPE
    A2PL3D('PAIR_FRAME_447',#602,#604,#603 );
    /* placement of pair relative to the 2nd link of joint*/
    #605 = &SCOPE    /* A2PL3D */
      #606 = CRTPNT('point_606',(0.0000,0.00025,0.00010 ) );
      #607 = DRCTN('dir_607',(0.00000,-1.00000,0.00000 ) );
      /* Y = DRCTN('dir6081',(0.00000,0.00000,1.00000)); */
      #608 = DRCTN('dir_608',(-1.00000,0.00000,0.00000 ) );
    ENDSCOPE
```

```
A2PL3D('PAIR_FRAME_452',#606,#608,#607 );
/* kinematic pair, value, and range */
#609 = PRSPR('pair_609','prismatic',#601,#605,#600 );
#610 = PRPRVL(#609,0.000000 );
#611 = PRPRRN(#609,-10.000000,10.000000 );
#612 = PRACT(#609,'actuator_612' );
/*    */
/* Kinematic Joint  KIN_JOINT_461 : */
#613 = KNMJNT(#597,#598 );
/* placement of pair relative to first  link of joint */
#614 = &SCOPE    /* A2PL3D */
  #615 = CRTPNT('point_615',(0.00040,0.00025,0.00081) );
  #616 = DRCTN('dir_616',(0.00000,0.00000,-1.00000) );
  /* Y = DRCTN('dir6171',(1.00000,0.00000,0.00000) ); */
  #617 = DRCTN('dir_617',(0.00000,-1.00000,0.00000 ) );
ENDSCOPE
A2PL3D('PAIR_FRAME_462',#615,#617,#616 );
/* placement of pair relative to the 2nd link of joint*/
#618 = &SCOPE    /* A2PL3D */
  #619 = CRTPNT('point_619',(0.00040,0.000251,0.00081));
  #620 = DRCTN('dir_620',(0.00000,0.00000,-1.000000));
  /* Y = DRCTN('dir6211',(1.00000,0.00000,0.00000)); */
  #621 = DRCTN('dir_621',(0.00000,-1.00000,0.00000));
ENDSCOPE
A2PL3D('PAIR_FRAME_467',#619,#621,#620 );
/* kinematic pair, value, and range */
#622 = RVLPR('pair_622','revolute',#614,#618,#613 );
#623 = RVPRVL(#622,0.000000 );
#624 = RVPRRN(#622,-6.283185,6.283185 );
#625 = PRACT(#622,'actuator_625' );
/*    */
/* Kinematic Joint  KIN_JOINT_476 : */
#626 = KNMJNT(#598,#599 );
/* placement of pair relative to first  link of joint */
#627 = &SCOPE    /* A2PL3D */
  #628 = CRTPNT('point_628',(0.00073,0.00025,0.00114));
  #629 = DRCTN('dir_629',(0.00000,0.00000,-1.00000));
  /* Y = DRCTN('dir6301',(1.00000,0.00000,0.00000)); */
  #630 = DRCTN('dir_630',(0.00000,-1.00000,0.00000));
ENDSCOPE
A2PL3D('PAIR_FRAME_477',#628,#630,#629 );
/* placement of pair relative to the 2nd link of joint*/
#631 = &SCOPE    /* A2PL3D */
  #632 = CRTPNT('point_632',(0.00073,0.00025,0.00114));
  #633 = DRCTN('dir_633',(0.00000,0.00000,-1.00000));
  /* Y = DRCTN('dir6341',(1.00000,0.00000,0.00000)); */
  #634 = DRCTN('dir_634',(0.00000,-1.00000,0.00000));
ENDSCOPE
A2PL3D('PAIR_FRAME_482',#632,#634,#633 );
/* kinematic pair, value, and range */
#635 = RVLPR('pair_635','revolute',#627,#631,#626 );
#636 = RVPRVL(#635,0.000000 );
#637 = RVPRRN(#635,-6.283185,6.283185 );
#638 = PRACT(#635,'actuator_638' );
/*    */
/* Link-Representation of LINK #596 */
```

```
    #639 = (GMRPCN(3 )GUAC((#100,#101 ))
        RPRCNT('LNK','Context for link-rep 640' ) );
    /* additional frame */
    #640 = &SCOPE    /* A2PL3D */
      #641 = CRTPNT('point_641',(0.0000,0.25000,0.10000 ) );
      #642 = DRCTN('dir_642',(0.00000,1.00000,0.00000 ) );
      /* Y = DRCTN('dir6431',(0.00000,0.00000,1.00000)); */
      #643 = DRCTN('dir_643',(1.00000,0.00000,0.00000 ) );
    ENDSCOPE
    A2PL3D('A2PL3D_491',#641,#643,#642 );
    #644 = KNLNRP('KIN_LINK_442',(#601,#640 ),#639 );
    #645 = KLRR(#596,#644 );
    #646 = FBSR('Shape_of_LINK_#596',(#161 ),#639 );
    #647 = (KLRA(* )RPRRLT('Rep_Rel_647',
        'Shape to Link_rep',#644,#646 )SHRPRL( ) );
    #648 = SHPASP('shape_aspect_648',
        'Connection to PRDFSH:',#585,.U. );
    #649 = PRPDFN('property_649',
        'Shape definition to shape aspect:',#648 );
    #650 = SHDFRP(#649,#646 );
    /*    */
    /* Link-Representation of LINK #597 */
    #651 = (GMRPCN(3 )GUAC((#100,#101 ))
        RPRCNT('LNK','Context for link-rep 652' ) );
    ...
    #661 = KLRR(#597,#660 );
    #662 = FBSR('Shape_of_LINK_#597',(#357 ),#651 );
    ...
    /* Link-Representation of LINK #598 */
    #667 = (GMRPCN(3 )GUAC((#100,#101 ))
       RPRCNT('LNK','Context for link-rep 668' ) );
    ...
    #677 = KLRR(#598,#676 );
    #678 = FBSR('Shape_of_LINK_#598',(#558 ),#667 );
    ...
    /* Link-Representation of LINK #599 */
    #683 = (GMRPCN(3 )GUAC((#100,#101 ))
       RPRCNT('LNK','Context for link-rep 684' ) );
    ...
    #689 = KLRR(#599,#688 );
    #690 = FBSR('Shape_of_LINK_#599',(#571 ),#683 );
    ...
    /*    */
    #695 = KNMSTR((#600,#613,#626 ) );
    #696 = MCBSPL('base of newenv_M441',
          'base_attachment',#588,*,#590,#595,* );
  ENDSCOPE
  MCHNSM(#695,#596,#587 );
  /* relation between property-def. and ground-
        representation */
  #697 = KPRR(#587,#588 );
ENDSCOPE
KNPRDF('property-def_587',
      'kinematic model and ground',#584,#586 );
ENDSEC;
END-ISO-10303-21;
```

Example STEP File Including Dynamics Information

```
ISO-10303-21;
HEADER;
FILE_DESCRIPTION( ( 'One bar pendulum' ), '1' );
FILE_NAME( 'o2.stp', '1995-11-28 T15:50:53',
           ( 'Sven Haas' ),
           ( 'Forschungszentrum Karlsruhe GmbH',
             '- Technik und Umwelt -',
             'HVT-HT',
             'Bau 691',
             'Postfach 3640',
             'D-76021 Karlsruhe',
             'Germany',
             'Telefon +49 7247 82-4213',
             'Telefax +49 7247 82-4795',
             'e-mail: Sven.Haas@hvt-ht.fzk.de',
             '=========================================' ),
             'robdb, RobDB 4.00',
             'o1',
             'FZK/HVT-HT%Sven%Haas' );
FILE_SCHEMA( ( 'IR_APPLICATION_SCHEMA_CC3' ) );
ENDSEC;
DATA;
/*    */
/* *** Global Units *** */
#100 = (LNGUNT( )NMDUNT(* )SUNT($,.METRE. ) );
#101 = (NMDUNT(* )PLANUN( )SUNT($,.RADIAN. ) );
#102 = (NMDUNT(* )SUNT($,.SECOND. )TMUNT( ) );
#103 = (MSSUNT( )NMDUNT(* )SUNT(.KILO.,.GRAM. ) );
#104 = (FORCE_UNIT( )NMDUNT(* )SUNT($,.NEWTON. ) );
#105 = DRUNEL(#100,1.000000 );
#106 = DRUNEL(#100,2.000000 );
#107 = DRUNEL(#101,1.000000 );
#108 = DRUNEL(#101,-1.000000 );
#109 = DRUNEL(#102,1.000000 );
#110 = DRUNEL(#102,-1.000000 );
#111 = DRUNEL(#102,-2.000000 );
#112 = DRUNEL(#103,1.000000 );
#113 = DRUNEL(#104,1.000000 );
#114 = ACCELERATION_UNIT((#105,#111 ) );
#115 = ANGULAR_VELOCITY_UNIT((#107,#110 ) );
#116 = ANGULAR_ACCELERATION_UNIT((#107,#111 ) );
#117 = MOMENT_OF_INERTIA_UNIT((#112,#106 ) );
#118 = TORQUE_UNIT((#105,#113 ) );
#119 = ROTATIONAL_DAMPING_COEFFICIENT_UNIT((#113,#105,#109,
    #110 ));
/*    */
/*    */
/* *** PRODUCT INFORMATION *** */
#120 = APPCNT(
    'InterRob model transfer / RobDB V2.0 / S1-G4-K1-K3' );
#121 = APPRDF('DRAFT_PROPOSAL','IR_APPLICATION_SCHEMA',1995,
    #120 );
#122 = PRDCNT('InterRob_model',#120,'Mechanical' );
```

```
#123 = PRDCT('OBAR','o1','***-sample-ADAMS-file',(#122 ) );
#124 = PRDFFR('VERS01','first_version',#123 );
#125 = PRPC('ROBOT','Modelled for InterRob',(#123 ) );
/*    */
/* *** workcell definitions for the kinematic model *** */
#126 = PRDFCN('Kinematic Model',#120,'Undefined' );
#127 = PRDDFN('Workcell_127','Kinematic model and ground',
    #124,#126 );
#128 = CHROBJ('environment_128',
    'The ground of the kinematic model' );
/* now comes the kinematic tree .... */
/*    */
/* *** The Kinematic Property Definition *** */
#129 = &SCOPE    /* KNPRDF */
 /*    */
 /* *** The Kinematic Ground Definition *** */
  #130 = &SCOPE    /* KNGRRP */
    #131 = INERTIA_CONTEXT(
        'Context for the shape of the ground (#130)','WCS',3,
        (#100,#101,#102,#103,#104,#114,#115,#116,#117,
        #118,#119 ) );
    #132 = &SCOPE    /* CTO3 */
      #133 = DRCTN('dir_133',(1.00000,0.00000,0.00000));
      #134 = DRCTN('dir_134',(0.00000,1.00000,0.00000));
      #135 = DRCTN('dir_135',(0.00000,0.00000,1.00000));
      #136 = CRTPNT('point_136',(0.00000,0.00000,0.00000));
    ENDSCOPE
    CTO3('rep_item_132','cto3_132','trafo',#133,#134,#136,
        1.000000,#135 );
    /*  gravitational acceleration  */
    #137 = &SCOPE    /* GRAVITATIONAL_FIELD */
      #138 = DRCTN('gravity orientation',
            (0.000000,0.000000,-1.000000 ) );
    ENDSCOPE
    GRAVITATIONAL_FIELD('gravity 137 in ground 0',#138,
            9.806650 );
  ENDSCOPE    / #132 /
  KNGRRP('kin_ground_rep_130',(#132,#137 ),#131 );
  /*    */
  /* *** The Definition of Mechanism  OBAR_M1 *** */
  #139 = &SCOPE    /* MCHNSM */
    /* Kinematic Link  PART_1 : */
    #140 = KNMLNK( );
    /* Kinematic Link  PART_2 : */
    #141 = KNMLNK( );
    /*    */
    /* Kinematic Joint  JOINT_1 : */
    #142 = KNMJNT(#140,#141 );
    /* placement of pair relative to first  link of joint */
    #143 = &SCOPE    /* A2PL3D */
      #144 = CRTPNT('point_144',(0.00000,0.00000,0.00000));
      #145 = DRCTN('dir_145',(0.00000,0.00000,1.00000));
      /* Y = DRCTN('dir1461',(-1.00000,0.00000,0.00000)); */
      #146 = DRCTN('dir_146',(0.00000,-1.00000,0.00000));
    ENDSCOPE    / #146 /
    A2PL3D('MARKER_1',#144,#146,#145 );
```

```
/* placement of pair relative to the 2nd link of joint*/
#147 = &SCOPE     /* A2PL3D */
  #148 = CRTPNT('point_148',(1.00000,0.00000,0.00000));
  #149 = DRCTN('dir_149',(1.00000,0.00000,0.00000));
  /* Y = DRCTN('dir1501',(0.00000,0.00000,1.00000));*/
  #150 = DRCTN('dir_150',(0.00000,-1.00000,0.00000));
ENDSCOPE
A2PL3D('MARKER_2',#148,#150,#149 );
/* kinematic pair, value, and range */
#151 = RVLPR('pair_151','revolute',#143,#147,#142 );
#152 = RVPRVL(#151,0.000000 );
#153 = RVPRRN(#151,-6.283185,6.283185 );
#154 = PRACT(#151,'actuator_154' );
/*     */
/* Link-Representation of LINK #140 */
#155 = INERTIA_CONTEXT('Context for link-rep 4','LNK',3,
    (#100,#101,#102,#103,#104,#114,#115,#116,#117,
      #118,#119 ) );
/* additional frame */
#156 = &SCOPE     /* A2PL3D */
  #157 = CRTPNT('point_157',(0.49437,0.00000,0.00000));
  #158 = DRCTN('dir_158',(1.00000,0.00000,0.00000));
  /* Y = DRCTN('dir1591',(0.00000,1.00000,0.00000)); */
  #159 = DRCTN('dir_159',(0.00000,0.00000,1.00000));
ENDSCOPE
A2PL3D('MARKER_11',#157,#159,#158 );
/* additional frame */
#160 = &SCOPE     /* A2PL3D */
  #161 = CRTPNT('point_161',(0.49437,0.00000,0.00000));
  #162 = DRCTN('dir_162',(1.00000,0.00000,0.00000));
  /* Y = DRCTN('dir1631',(0.00000,1.00000,0.00000)); */
  #163 = DRCTN('dir_163',(0.00000,0.00000,1.00000));
ENDSCOPE
A2PL3D('MARKER_12',#161,#163,#162 );
/* additional frame */
#164 = &SCOPE     /* A2PL3D */
  #165 = CRTPNT('point_165',(0.00000,0.00000,0.00000));
  #166 = DRCTN('dir_166',(1.00000,0.00000,0.00000));
  /* Y = DRCTN('dir1671',(0.00000,1.00000,0.00000)); */
  #167 = DRCTN('dir_167',(0.00000,0.00000,1.00000));
ENDSCOPE
A2PL3D('MARKER_10',#165,#167,#166 );
#168 = KNLNRP('PART_1',(#156,#160,#164,#143 ),#155 );
#169 = KLRR(#140,#168 );
/*     */
/* Link-Representation of LINK #141 */
#170 = INERTIA_CONTEXT('Context for link-rep 4','LNK',3,
    (#100,#101,#102,#103,#104,#114,#115,#116,#117,
      #118,#119 ) );
/* additional frame */
#171 = &SCOPE     /* A2PL3D */
  #172 = CRTPNT('point_172',(0.00000,0.00000,0.00000));
  #173 = DRCTN('dir_173',(1.00000,0.00000,0.00000));
  /* Y = DRCTN('dir1741',(0.00000,1.00000,0.00000)); */
  #174 = DRCTN('dir_174',(0.00000,0.00000,1.00000));
ENDSCOPE
```

```
    A2PL3D('MARKER_13',#172,#174,#173 );
    /* additional frame */
    #175 = &SCOPE     /* A2PL3D */
      #176 = CRTPNT('point_176',(0.00000,0.00000,0.00000));
      #177 = DRCTN('dir_177',(1.00000,0.00000,0.00000));
      /* Y = DRCTN('dir1781',(0.00000,0.00000,1.00000)); */
      #178 = DRCTN('dir_178',(0.00000,-1.00000,0.00000));
    ENDSCOPE
    A2PL3D('MARKER_3',#176,#178,#177 );
    #179 = KNLNRP('PART_2',(#171,#147,#175 ),#170 );
    #180 = KLRR(#141,#179 );
    /* mass of */
    /* PART_2 */
    #181 = &SCOPE     /* MASS_PROPERTIES_ITEM */
      #182 = CRTPNT('centre of mass_182',
        (0.500000,0.000000,0.000000 ) );
      #183 = DRCTN('z_axis_183',(0.00000,0.00000,1.00000));
      #184 = DRCTN('x_axis_184',(1.00000,0.00000,0.00000));
      #185 = A2PL3D('inertia_frame_185',#182,#183,#184 );
      #186 = CRTPNT('centre of gravity_186',
        (0.500000,0.000000,0.000000 ) );
    ENDSCOPE
    MASS_PROPERTIES_ITEM('mass of #141',10.000000,
        (0.01667,0.84167,0.84167,0.00000,0.00000,0.00000),
        #186,#185 );
    #187 = MASS_PROPERTIES_REPRESENTATION('m_PART_2',
        (#181 ),#170 );
    #188 = (KLRA(* )MASS_PROPERTIES_TO_PART_ASSOCIATION( )
        RPRRLT('rep-rel #188', 'link-mass',#179,#187 ) );
    /*    */
    #189 = KNMSTR((#142 ) );
    #190 = MCBSPL('base of
OBAR_M1','base_attachment',#130,*,#132,#139,* );
  ENDSCOPE     / #146,#151,#155,#164,#168,#171,#179,#187 /
  MCHNSM(#189,#140,#129 );
  /* relation between property-def.
     and ground-representation */
  #191 = KPRR(#129,#130 );
ENDSCOPE     / #146,#151,#155,#164,#168,#171,#179,#187 /
KNPRDF('property-def_129',
    'kinematic model and ground',#127,#128 );
#192 = &SCOPE     /* DYNAMIC_ANALYSIS_DATA */
  #193 = &SCOPE     /* MULTI_BODY_SYSTEM */
    /* initial states */
    #194 = (DISPLACEMENT_STATE_ELEMENT( )PRVL(#151 )
        PAIR_VALUE_STATE( )
        RPRITM('initial value of pair 151' )
        RVPRVL(0.000000 )STATE_ELEMENT( ) );
    #195 = ANGULAR_VELOCITY_ABOUT_DIRECTION(
        'initvel in pair 151',#146,0.000000 );
    /* evaluators and interactions */
    #196 = SPATIAL_LOAD_EVALUATOR(
        'SLE-MARKER_1/MARKER_2 in pair 151' );
    #197 = PAIR_VALUE_EVALUATOR(
        'PVE-MARKER_1/MARKER_2 in pair 151' );
    #198 = DYNAMIC_INTERACTORS_REPRESENTATION(
```

```
          'MARKER_1 to MARKER_2 in pair 151',
        (#194,#195,#196,#197 ),#155 );
    #199 = (DYNAMIC_INTERACTOR_ANCHOR_ASSOCIATION( )KLRA( )
          RPRRLT('rep-rel 199', 'interactors',#168,#198 ) );
    #200 = DYNAMIC_INTERACTION('inter-12/15',
          'Interaction in pair #151',( ),(#194,
          #195 ),(#196,#197 ),( ) );
    #201 = KINEMATICALLY_BOUNDED_PARTS_RELATIONSHIP(
          'inter-12/15-p',
          'Interaction in pair #151',*,*,#200,*,*,#151 );
    #202 = TRANSFORM_YPR_EVALUATOR(
          'TYE-MARKER_10/MARKER_13');
    #203 = DYNAMIC_INTERACTORS_REPRESENTATION(
          'MARKER_10 to MARKER_13',(#202 ),#155 );
    #204 = (DYNAMIC_INTERACTOR_ANCHOR_ASSOCIATION( )KLRA( )
          RPRRLT('rep-rel 204', 'interactors',#168,#203 ) );
    #205 = DYNAMIC_INTERACTION('inter-11/16',
          'Interaction MARKER_10/MARKER_13',( ),
          ( ),(#202 ),( ) );
    #206 = DYNAMIC_PARTS_RELATIONSHIP('inter-11/16-p',
          'Interaction MARKER_10/MARKER_13',
          #168,#179,#205,#164,#171 );
    #207 = TRANSFORM_YPR_EVALUATOR(
          'TYE-MARKER_10/MARKER_10');
    #208 = DYNAMIC_INTERACTORS_REPRESENTATION(
          'MARKER_10 to MARKER_10',(#207 ),#155 );
    #209 = (DYNAMIC_INTERACTOR_ANCHOR_ASSOCIATION( )KLRA( )
          RPRRLT('rep-rel 209',
          'interactors',#168,#208 ) );
    #210 = DYNAMIC_INTERACTION('inter-11/11',
          'Interaction MARKER_10/MARKER_10',( ),
          ( ),(#207 ),( ) );
    #211 = DYNAMIC_PARTS_RELATIONSHIP('inter-11/11-p',
          'Interaction MARKER_10/MARKER_10',
          #168,#168,#210,#164,#164 );
  ENDSCOPE    / #194,#195,#196,#197,#202,#207 /
  MULTI_BODY_SYSTEM((#187 ),(#201,#206,#211 ),( ) );
  #212 = &SCOPE    /* DYNAMIC_ANALYSIS_TASK */
    #213 = TMWU(TIME_MEASURE(0.000000 ),#102 );
    #214 = TMWU(TIME_MEASURE(10.000000 ),#102 );
    #215 = TMWU(TIME_MEASURE(1.000000E-02 ),#102 );
    #216 = TMWU(TIME_MEASURE(1.000000 ),#102 );
    #217 = CNFDFN((#194 ),#213 );
    #218 = VELOCITY_STATE_SET(#213,(#195 ) );
    #219 = DYNAMIC_OUTPUT_REQUEST((#196,#197,#202,#207 ) );
  ENDSCOPE
  DYNAMIC_ANALYSIS_TASK(#192,#193,( ),#213,#214,#215,#216,
        1.000000E-2,(#217 ),(#218 ),#219 );
ENDSCOPE
DYNAMIC_ANALYSIS_DATA(#129,(#193 ) );
ENDSEC;
END-ISO-10303-21;
```

Example STEP File Including Dynamics and Control.

T. Sørensen
Danmarks Tekniske Universitet, Instituttet for Styreteknik
Bygning 424, DK-2800 Lyngby, Denmark

```
ISO-10303-21;
HEADER;
/*----------------------------------------
 * Exchange File generated by ST-DEVELOPER v1.4
 * Conforms to ISO 10303-21
 */
FILE_DESCRIPTION (('Joint 1 of a robot with manipulator
dynamics and PI (speed) plus P (position) controllers  '),
'1');
FILE_NAME ('robot.step', '1995-11-21T16:11:35+01:00',
('Torben Sorensen'), ('IPU Control Engineering section Build.
424 1. floor',
 'Technical University of Denmark',
 'DK-2800 Lyngby'),
 'ST-DEVELOPER v1.4', '', '');
FILE_SCHEMA (('CONTROLLER_SCHEMA'));
ENDSEC;

DATA;
#10 = LINEAR_STATE_SPACE_SYSTEM (.T., 3, 2, 3,
((0., 1., 0.),(-14567., -384., 80.), (-22800., -600., 0.)),
((0., 0.), (14567., 383.), (22800., 600.)),
((1., 0., 0.), (0., 1., 0.), (-14567., -384., 800.)),
((0., 0.), (0., 0.), (14567., 383.)));
ENDSEC;
END-ISO-10303-21;
```

Example of a HICADEC/P STEP file

The following file is an example of a STEP file for transferring information on pipe geometry from the CAD system HICADEC/P to the product model database. The STEP file is according to an EXPRESS schema developed for this purpose in InterRob, OSS_piping_model schema.

```
ISO-10303-21;
HEADER;
FILE_NAME('pipe_4.stp','1994-07-15 T11:27:51', ('Frode
      Hoegberg'),('SINTEF'),'','manual_new','');
FILE_DESCRIPTION(('Simple piping model for OSS'),'1');
FILE_SCHEMA(('OSS_piping_model'));
ENDSEC;
DATA;

/*The following entities represent the first flange in NP034001*/

#10 = MEASURE_WITH_LENGTH_UNIT(0.00,.METRE.);
#20 = MEASURE_WITH_LENGTH_UNIT(0.00,.METRE.);
#30 = MEASURE_WITH_LENGTH_UNIT(0.00,.METRE.);
#40 = OSS_CARTESIAN_POINT((#10,#30,#20));
#50 = MEASURE_WITH_LENGTH_UNIT(0.00,.METRE.);
#60 = MEASURE_WITH_LENGTH_UNIT(0.032,.METRE.);
#70 = MEASURE_WITH_LENGTH_UNIT(0.00,.METRE.);
#80 = OSS_CARTESIAN_POINT((#60,#70,#50));
#90 = CENTER_LINE((#80,#40));
#100 = MEASURE_WITH_LENGTH_UNIT(565.0,.MILLIMETRE.);
#110 = OPENING(.HOT_GALVANIZED.,.HOT_GALVANIZED.,#100,#130);
#120 = MEASURE_WITH_WEIGHT_UNIT(100.0,.KILOGRAM.);
#130 = MEASURE_WITH_LENGTH_UNIT(78.5,.MILLIMETRE.);
#140 = FLANGE((#1100), 'flangecode1', 'base_pipe_fl1', 'FLANGE',
      34001, .ST37_2., #120, #110, #110, #90);

/*The following entities represent the base pipe in NP034001*/

#150 = MEASURE_WITH_LENGTH_UNIT(0.00,.METRE.);
#160 = MEASURE_WITH_LENGTH_UNIT(1.010,.METRE.);
#170 = MEASURE_WITH_LENGTH_UNIT(0.00,.METRE.);
#180 = OSS_CARTESIAN_POINT((#160,#150,#170));
#190 = CENTER_LINE((#180,#80));
#200 = MEASURE_WITH_LENGTH_UNIT(406.8,.MILLIMETRE.);
#210 = OPENING(.HOT_GALVANIZED.,.HOT_GALVANIZED.,#200,#230);
#220 = MEASURE_WITH_WEIGHT_UNIT(100.0,.KILOGRAM.);
#230 = MEASURE_WITH_LENGTH_UNIT(8.80,.MILLIMETRE.);
#240 = PIPE((#1100), 'pipecode1', 'base_pipe', 'PIPE', 34001,
      .ST37_2., #220, #210, #210, #190);


#250 = MEASURE_WITH_LENGTH_UNIT(0.610,.METRE.);
#260 = MEASURE_WITH_LENGTH_UNIT(1.620,.METRE.);
#270 = MEASURE_WITH_LENGTH_UNIT(0.00,.METRE.);
#280 = OSS_CARTESIAN_POINT((#260,#250,#270));

/*The following entities represent the connection pipe in
      NP034001*/

#390 = MEASURE_WITH_LENGTH_UNIT(0.00,.METRE.);
```

128   F. Høgberg et al.

```
#400 = MEASURE_WITH_LENGTH_UNIT(0.600,.METRE.);
#410 = MEASURE_WITH_LENGTH_UNIT(0.00,.METRE.);
#420 = OSS_CARTESIAN_POINT((#400,#390,#410));
#430 = MEASURE_WITH_LENGTH_UNIT(0.00,.METRE.);
#440 = MEASURE_WITH_LENGTH_UNIT(0.600,.METRE.);
#450 = MEASURE_WITH_LENGTH_UNIT(0.429,.METRE.);
#460 = OSS_CARTESIAN_POINT((#440,#430,#450));
#470 = CENTER_LINE((#460,#420));
#480 = MEASURE_WITH_LENGTH_UNIT(273.0,.MILLIMETRE.);
#490 = OPENING(.HOT_GALVANIZED.,.HOT_GALVANIZED.,#480,#510);
#500 = MEASURE_WITH_WEIGHT_UNIT(100.0,.KILOGRAM.);
#510 = MEASURE_WITH_LENGTH_UNIT(7.10,.MILLIMETRE.);
#520 = PIPE((#1100), 'pipecode1', 'conn_pipe', 'PIPE', 34001,
       .ST37_2., #500, #490, #490, #470);

/*The following entities represent the bender in NP034001*/

#530 = MEASURE_WITH_LENGTH_UNIT(0.00,.METRE.);
#540 = MEASURE_WITH_LENGTH_UNIT(1.620,.METRE.);
#550 = MEASURE_WITH_LENGTH_UNIT(0.00,.METRE.);
#560 = OSS_CARTESIAN_POINT((#540,#550,#530));
#570 = CENTER_LINE((#180,#280,#560));
#580 = MEASURE_WITH_LENGTH_UNIT(406.8,.MILLIMETRE.);
#590 = OPENING(.HOT_GALVANIZED.,.HOT_GALVANIZED.,#580,#610);
#600 = MEASURE_WITH_WEIGHT_UNIT(100.0,.KILOGRAM.);
#610 = MEASURE_WITH_LENGTH_UNIT(8.80,.MILLIMETRE.);
#620 = BENDER((#1100), 'bender_code_1', 'bend_pipe', 'PIPE', 34001,
       .ST37_2., #600, #590, #590, #570);

/*The following entities represent the flange on the connection
       pipe in NP034001*/

#630 = MEASURE_WITH_LENGTH_UNIT(0.00,.METRE.);
#640 = MEASURE_WITH_LENGTH_UNIT(0.600,.METRE.);
#650 = MEASURE_WITH_LENGTH_UNIT(0.455,.METRE.);
#660 = OSS_CARTESIAN_POINT((#640,#630,#650));
#670 = CENTER_LINE((#660,#460));
#680 = MEASURE_WITH_LENGTH_UNIT(395.0,.MILLIMETRE.);
#690 = OPENING(.HOT_GALVANIZED.,.HOT_GALVANIZED.,#680,#710);
#700 = MEASURE_WITH_WEIGHT_UNIT(100.0,.KILOGRAM.);
#710 = MEASURE_WITH_LENGTH_UNIT(60.5,.MILLIMETRE.);
#720 = FLANGE((#1100), 'flangecode2',' conn_pipe_fl1',' FLANGE',
       34001, .ST37_2.,# 700, #690, #690, #670);

#1100 = OSS_PRODUCT('test_product_4');

ENDSEC;
END-ISO-10303-21;•
```

Example STEP file conversion of welding specific data

The following example serves to illustrate the data route from the InterRob Database, developed under Work Package 3 for use at OSS, to the GRASP simulation and off-line programming system. The data conforms to the additional schemas for Weld Process and Weld Data (A1 and A3) developed outside the scope of the InterRob STEP specification.

Both example files below make use of ellipses to indicate the editing of repetitious common entries.

STEP file from the InterRob database
(weld data and process schemas, A1 and A3)

```
ISO-10303-21;
HEADER;
/*----------------------------------------
 * Exchange File generated by ST-DEVELOPER v1.4
 * Conforms to ISO 10303-21
 */
FILE_DESCRIPTION ((''), '1');
FILE_NAME ('export.cc5', '1995-11-02T18:57:49+01:00', (''), (''),
        'ST-DEVELOPER v1.4', '', '');
FILE_SCHEMA (('IR_APPLICATION_SCHEMA'));
ENDSEC;

DATA;
#10 = WELD_POINT_WEAVE_DATA (1.5, 2.29999995231628,
        2.29999995231628, 0., 3.14159265358979, 35.);
#20 = WELD_POINT_PROCESS_DATA ((#30, #40), 0., #23990,
        1.47759997844696, #10, .T., .T., -1.37550624943896);
#30 = VOLTAGE_CHANNEL (1, .IS_REAL., 3.44000005722046);
#40 = VOLTAGE_CHANNEL (3, .IS_REAL., 4.21999979019165);
#50 = WELD_POINT_WEAVE_DATA (1.5, 2.29999995231628,
        2.29999995231628, 0., 3.14159265358979, 35.);
#60 = WELD_POINT_PROCESS_DATA ((#70, #80), 0., #24000,
        1.47000002861023, #50, .T., .T., -1.36926282671227);
#70 = VOLTAGE_CHANNEL (1, .IS_REAL., 3.44000005722046);
#80 = VOLTAGE_CHANNEL (3, .IS_REAL., 4.21999979019165);
...
#2850 = WELD_POINT_WEAVE_DATA (1.5, 1.77999997138977,
        1.77999997138977, 0., 3.14159265358979, 35.);
#2860 = WELD_POINT_PROCESS_DATA ((#2870, #2880), 0., #24700,
        2.23379993438721, #2850, .T., .T., -1.36926282671227);
#2870 = VOLTAGE_CHANNEL (1, .IS_REAL., 3.44000005722046);
#2880 = VOLTAGE_CHANNEL (3, .IS_REAL., 4.21999979019165);
#2890 = WELD_PROCESS_DATA ('weld_process_data',
        'weld_process_data', 'Weld process data', #37230, #37230,
        ((#20, #60, #100, #140, #180, #220, #260, #300, #340, #380,
        #420, #460, #500, #540, #580, #620, #660, #700, #740, #780,
        #820, #860, #900, #940, #980, #1020, #1060, #1100, #1140,
        #1180, #1220, #1260, #1300, #1340, #1380, #1420, #1460,
        #1500, #1540, #1580, #1620, #1660, #1700, #1740, #1780,
        #1820, #1860, #1900, #1940, #1980, #2020, #2060, #2100,
        #2140, #2180, #2220, #2260, #2300, #2340, #2380, #2420,
```

```
         #2460, #2500, #2540, #2580, #2620, #2660, #2700, #2740,
         #2780, #2820, #2860), (#2910, #2950, #2990, #3030, #3070,
         #3110, #3150, #3190, #3230, #3270, #3310, #3350, #3390,
         #3430, #3470, #3510, #3550, #3590, #3630, #3670, #3710,
         #3750, #3790, #3830, #3870, #3910, #3950, #3990, #4030,
         #4070, #4110, #4150, #4190, #4230, #4270, #4310, #4350,
         #4390, #4430, #4470, #4510, #4550, #4590, #4630, #4670,
         #4710, #4750, #4790, #4830, #4870, #4910, #4950, #4990,
         #5030, #5070, #5110, #5150, #5190, #5230, #5270, #5310,
         #5350, #5390, #5430, #5470, #5510, #5550, #5590, #5630,
         #5670, #5710, #5750), (#5790, #5830, #5870, #5910, #5950,
         #5990, #6030, #6070, #6110, #6150, #6190, #6230, #6270,
         #6310, #6350, #6390, #6430, #6470, #6510, #6550, #6590,
         #6630, #6670, #6710, #6750, #6790, #6830, #6870, #6910,
         #6950, #6990, #7030, #7070, #7110, #7150, #7190, #7230,
         #7270, #7310, #7350, #7390, #7430, #7470, #7510, #7550,
         #7590, #7630, #7670, #7710, #7750, #7790, #7830, #7870,
         #7910, #7950, #7990, #8030, #8070, #8110, #8150, #8190,
         #8230, #8270, #8310, #8350, #8390, #8430, #8470, #8510,
         #8550, #8590, #8630)));
#2900 = WELD_POINT_WEAVE_DATA (1.5, 2.29999995231628,
         2.29999995231628, 0., 3.14159265358979, 0.);
...
#8630 = WELD_POINT_PROCESS_DATA ((#8640, #8650), 0., #26140, 6.,
         #8620, .T., .T., -1.36926282671227);
#8640 = VOLTAGE_CHANNEL (1, .IS_REAL., 6.13000011444092);
#8650 = VOLTAGE_CHANNEL (3, .IS_REAL., 6.65000009536743);
#8660 = CARTESIAN_POINT ('point', (129.403614972387,
         168.928705698244, 0.));
...
#8790 = CARTESIAN_POINT ('point', (129.405261115911,
         156.926219592275, 0.));
#8800 = CROSS_SECTION_LINE ('branch_inner', (#8660, #8700),
         .PART1_INNER.);
...
#8850 = CROSS_SECTION_LINE ('main_outer', (#8770, #8780, #8790),
         .PART2_OUTER.);
#8860 = WELD_CROSS_SECTION (0., (#23990, #24710, #25430), (#8800,
         #8810, #8820, #8830, #8840, #8850), ());
#8870 = CARTESIAN_POINT ('point', (129.402214880593,
         169.334391984366, 0.));
...
#23970 = CROSS_SECTION_LINE ('main_outer', (#23890, #23900,
         #23910), .PART2_OUTER.);
#23980 = WELD_CROSS_SECTION (6.28318530717959, (#23990, #24710,
         #25430), (#23920, #23930, #23940, #23950, #23960, #23970),
         ());
#23990 = WELD_POINT ('point', (-2.11310005187988, 133.898696899414,
         157.491500854492), 1, #29810, 868.549279638691, 2., 0.,
         0.785398163397448, 0., -0.0523598775598299,
         0.436332312998582);
#24000 = WELD_POINT ('point', (-13.7782001495361, 133.211502075195,
         158.068206787109), 1, #26970, 11.6995464334088, 2., 0.,
         0.785398163397448, 0., -0.0523598775598299,
         0.436332312998582);
...
#26130 = WELD_POINT ('point', (24.9930992126465, 139.107406616211,
         155.931304931641), 3, #37070, 897.306944206721, 2., 0.,
         0.785398163397448, 0., -0.0523598775598299,
         0.436332312998582);
#26140 = WELD_POINT ('point', (12.5501003265381, 140.746994018555,
         154.719299316406), 3, #37110, 909.915886499434, 2., 0.,
```

```
      0.785398163397448, 0., -0.0523598775598299,
      0.436332312998582);
#26150 = B_SPLINE_CURVE_WITH_KNOTS ('weld_curve_b_spline_curve', 3,
      (#26160, #26170, #26180, #26190, #26200, #26210, #26220,
      #26230, #26240, #26250, #26260, #26270, #26280, #26290,
      #26300, #26310, #26320, #26330, #26340, #26350, #26360,
      #26370, #26380, #26390, #26400, #26410, #26420, #26430,
      #26440, #26450, #26460, #26470, #26480, #26490, #26500,
      #26510, #26520, #26530, #26540, #26550, #26560, #26570,
      #26580, #26590, #26600, #26610, #26620, #26630, #26640,
      #26650, #26660, #26670, #26680, #26690, #26700, #26710,
      #26720, #26730, #26740, #26750, #26760, #26770, #26780,
      #26790, #26800, #26810, #26820, #26830, #26840, #26850,
      #26860, #26870, #26880), .UNSPECIFIED., .F., .F., (4, 1, 1,
      1, 1, 1, 1, 1, 1, 1, 1, 1, 1, 1, 1, 1, 1, 1, 1, 1, 1, 1,
      1, 1, 1, 1, 1, 1, 1, 1, 1, 1, 1, 1, 1, 1, 1, 1, 1, 1, 1,
      1, 1, 1, 1, 1, 1, 1, 1, 1, 1, 1, 1, 1, 1, 1, 1, 1, 1, 1,
      1, 1, 1, 1, 1, 1, 1, 4), (0., 23.4503999841684,
      35.2946463820364, 47.2581821155551, 59.3483514596778,
      71.5542937454166, 83.8511176385901, 96.204484727576,
      108.576746186487, 120.930853817929, 133.23482081803,
      145.464100811451, 157.60351542652, 169.647753213221,
      181.60174300586, 193.478966806163, 205.300802598979,
      217.093403697826, 228.886024023718, 240.707895302512,
      252.585414028656, 264.539612459384, 276.58420021595,
      288.723947306559, 300.953607978366, 313.257963517894,
      325.612470616869, 337.984972460866, 350.338606815802,
      362.63548238491, 374.841423013511, 386.93136915722,
      398.894596870312, 410.738430681166, 422.489126549126,
      434.188571941237, 445.888118374646, 457.638971925406,
      469.483218323274, 481.446754056793, 493.536923400915,
      505.742865686654, 518.039689579827, 530.393056668813,
      542.765318127724, 555.119425759167, 567.423392759267,
      579.652672752689, 591.792087367758, 603.836325154459,
      615.790314947098, 627.667538747401, 639.489374540216,
      651.281975639063, 663.07839078192, 674.907847500647,
      686.788809965177, 698.739856892262, 710.780225538461,
      722.914292521175, 735.13679256129, 747.436476269568,
      759.787625478702, 772.158954655166, 784.515298345952,
      796.818740719782, 809.040529084267, 821.151092155665,
      833.138251840496, 845.019142839174, 868.549279638691),
      .UNSPECIFIED.);
#26160 = CARTESIAN_POINT ('.Cart pt.', (-2.11310005187988,
      133.898696899414, 157.491500854492));
#26170 = CARTESIAN_POINT ('.Cart pt.', (-9.9349299048859,
      133.778088211395, 157.58718992526));
...
#26870 = CARTESIAN_POINT ('.Cart pt.', (5.71398525676836,
      134.028155060661, 156.936586560863));
#26880 = CARTESIAN_POINT ('.Cart pt.', (-2.11310005187988,
      133.898696899414, 157.491500854492));
#26890 = WELD_CURVE (#26150, 1, (#23990, #24000, #24010, #24020,
      #24030, #24040, #24050, #24060, #24070, #24080, #24090,
      #24100, #24110, #24120, #24130, #24140, #24150, #24160,
      #24170, #24180, #24190, #24200, #24210, #24220, #24230,
      #24240, #24250, #24260, #24270, #24280, #24290, #24300,
      #24310, #24320, #24330, #24340, #24350, #24360, #24370,
      #24380, #24390, #24400, #24410, #24420, #24430, #24440,
      #24450, #24460, #24470, #24480, #24490, #24500, #24510,
      #24520, #24530, #24540, #24550, #24560, #24570, #24580,
      #24590, #24600, #24610, #24620, #24630, #24640, #24650,
      #24660, #24670, #24680, #24690, #24700, #23990));
```

```
#26900 = CARTESIAN_POINT ('point', (-2.11310005187988,
     133.898696899414, 157.491500854492));
#26910 = DIRECTION ('axis_dir', (-0.0186728681993536,
     0.939396652197142, -0.342323317689596));
#26920 = DIRECTION ('axis_dir', (-0.999806345432029, -
     0.0154165117193905, 0.0122312223532867));
#26930 = AXIS2_PLACEMENT_3D ('bisectorframe', #26900, #26910,
     #26920);
#26940 = CARTESIAN_POINT ('point', (-13.7782001495361,
     133.211502075195, 158.068206787109));
#26950 = DIRECTION ('axis_dir', (-0.123587971047686,
     0.940077279223934, -0.317774641057531));
#26960 = DIRECTION ('axis_dir', (-0.991178440221944, -
     0.101496461487605, 0.0852277416378636));
...
#29770 = AXIS2_PLACEMENT_3D ('bisectorframe', #29740, #29750,
     #29760);
#29780 = CARTESIAN_POINT ('point', (-2.11310005187988,
     133.898696899414, 157.491500854492));
#29790 = DIRECTION ('axis_dir', (-0.0396423404823882,
     0.939606915288667, -0.339951951873783));
#29800 = DIRECTION ('axis_dir', (-0.997360536125632, -
     0.0164961101119417, 0.070709542003661));
#29810 = AXIS2_PLACEMENT_3D ('bisectorframe', #29780, #29790,
     #29800);
#29820 = B_SPLINE_CURVE_WITH_KNOTS ('weld_curve_b_spline_curve', 3,
     (#29830, #29840, #29850,
...
#37180 = WELD_DESIGN_DATA (('BASE_PIPE', 'CONN_PIPE'), 2.,
     0.785398163397448, 0.1);
#37190 = WELD_SEAM_DATA ('BASEPIPE_BENDPIPE_WELDDATA', (), (),
     #37200);
#37200 = WELD_DESIGN_DATA (('BASE_PIPE', 'BEND_PIPE'), 2.,
     0.523598775598299, 0.1);
#37210 = PRODUCT_DEFINITION_FORMATION ('PEH951102',
     'Triple_layer_weld', #37290);
#37220 = PRODUCT_DEFINITION_CONTEXT ('Geometry Model Shape',
     #37260, 'Undefined');
#37230 = PRODUCT_DEFINITION ('CC1', 'IPPG Pipe Composite', #37210,
     #37220);
#37240 = PRODUCT_DEFINITION_SHAPE ('Shape_property', 'Pipe assembly
     with groove geometry [and weld data [and weld process
     data]]', #37230);
#37250 = GEOMETRIC_REPRESENTATION_CONTEXT
     ('IPPG_geometric_context', 'Context for pipe assembly and
     weld shape', 3);
#37260 = APPLICATION_CONTEXT ('InterRob Pipe Program Generation
     System');
#37270 = APPLICATION_PROTOCOL_DEFINITION ('DRAFT PROPOSAL',
     'IR_APPLICATION_SCHEMA_CC2', 1995, #37260);
#37280 = PRODUCT_CONTEXT ('InterRob_model', #37260, 'Mechanical');
#37290 = PRODUCT ('test_product_4', '3D-solid/surface',
     'OSS_demo_pipe', (#37280));
#37300 = PRODUCT_RELATED_PRODUCT_CATEGORY ('PART', 'Spools for
     robot welding', (#37290));
ENDSEC;
END-ISO-10303-21;
```

The corresponding GRASP source file representation follows:

```
{*******************************************************************}
{                                                                 }
{ -> GRASP SOURCE FILE GENERATED BY                        <- }
{ -> STEP_TO_GRASP PREPROCESSOR VERSION  1.1_____ <- }
{    ~~~~~~~~~~~~~~~~~~~~~~~~~~~~~~~~~~~~~~~~~~~~~~~~~~~~~~~~~~    }
{                                                                 }
{*******************************************************************}

{ File Created 3/11/1995 16:1:41 }

{~~~~~~~~~~~~~~~~}
{ HEADER SECTION }
{~~~~~~~~~~~~~~~~}

{ File Name *}

{    File name        : export.cc5
{    Time Stamp       : 1995-11-02T18:57:49+01:00
{    Author           :
{    Organisation     :
{    Preproc Version  : ST-DEVELOPER v1.4
{    Original system  :
{    Authorisation    :

{* End of file identifer }

{ File Description *}

{    Description          :
{    Implementation Level : 1

{* End of file description }


{ Weld Templates }

weld_list_template LT2890_2890
length 7 speed_item 1 dwell_time_item 2
    item 1   type real    label 'Speed (mm/s)'
    item 2   type real    label 'Pause (s)'
    item 3   type logical label 'Weld on ?'
    item 4   type real    label 'Gravity angle (deg.)'
    item 5   type real    label 'Power   Ch.1  (V)'
    item 6   type real    label 'Voltage Ch.3  (V)'
    item 7   type real    label 'Pulse control (V)'
;

weld_weave_template WT2890_2890
length 8
    item 1   type logical label 'Weave on ?
    item 2   type real    label 'Amplitude left   (mm)'
    item 3   type real    label 'Amplitude right  (mm)'
    item 4   type real    label 'Angle left     (deg.)'
    item 5   type real    label 'Angle right    (deg.)'
    item 6   type real    label 'Frequency       (Hz)'
    item 7   type real    label 'Halt time       (%)'
    item 8   type name    label 'Aux. point defn. '
;
```

```
weld_parameter_list PL20_20
    template LT2890_2890
         1.478 0.000      TRUE        0.000       3.440 4.220 0.000
;
weld_weave_parameters WP20_20
    template WT2890_2890
         TRUE       2.300 2.300 0.000 180.000 1.500 35.000   bf_23990
;
weld_parameter_list PL60_60
    template LT2890_2890
         1.470 0.000      TRUE        0.000       3.440 4.220 0.000
;
weld_weave_parameters WP60_60
    template WT2890_2890
         TRUE       2.300 2.300 0.000 180.000 1.500 35.000   bf_24000
;
...

weld_weave_parameters WP8630_8630
    template WT2890_2890
         TRUE       1.780 1.780 0.000 180.000 1.500 0.000    bf_26140
;
{ B Spline Curve : entity #26150 }
CURVE NURB C26150_26150  DEGREE 3 NO_POINTS 73 NON_RATIONAL
NON_PERIODIC
FROM 0.0000 TO 868.5493
KNOTS
         0.0000        0.0000        0.0000        0.0000
         23.4504       35.2946       47.2582       59.3484
...
         809.0405      821.1511      833.1382      845.0192
         868.5493      868.5493      868.5493      868.5493
POINTS
-2.1131 133.8987 157.4915
-9.9349 133.7781 157.5872
...
5.7140 134.0282 156.9366
-2.1131 133.8987 157.4915
;

{ B Spline Curve : entity #29820 }
CURVE NURB C29820_29820 DEGREE 3 NO_POINTS 73 NON_RATIONAL
      NON_PERIODIC
FROM 0.0000 TO 875.5116
KNOTS
         0.0000
...
         875.5116
POINTS
0.0000 134.5000 150.2704
...
0.0000 134.5000 150.2704
;

{ B Spline Curve : entity #33490 }
CURVE NURB C33490_33490  DEGREE 3 NO_POINTS 73 NON_RATIONAL
      NON_PERIODIC
FROM 0.0000 TO 922.4810
KNOTS
         0.0000
...
         922.4810
```

```
POINTS
0.0000 141.2949 154.9984
...
0.0000 141.2949 154.9984
;

{ Weld Seam Data  }

set W37170_37170  = ;

{ Seam - weld curve : entity #26890 }
SEAM S37170_26890
   PRISM_SIDES 3 PRISM_RADIUS 1.0 VECTOR_LENGTH 20.0
    list_template LT2890_2890  weave_template WT2890_2890
   COMPONENT POLYLINE
0.0000 0.0000 0.0000
      BISECTOR_FRAME bf_23990s (euler 92.4159 109.8739 95.2751)
      VECTOR (euler -62.3807 76.3037 -101.7527)
            parameter_list PL20_20
            weave_parameters WP20_20
-11.6651 -0.6872 0.5767
      BISECTOR_FRAME bf_24000 (euler 97.4895 108.5284 96.2679)
      VECTOR (euler -57.5820 77.8515 -102.0228)
            parameter_list PL60_60
            weave_parameters WP60_60
...
11.7287 -0.3230 -0.2429
      BISECTOR_FRAME bf_24700 (euler 85.3590 108.6683 87.8320)
      VECTOR (euler -68.5258 75.1699 -94.9458)
            parameter_list PL2860_2860
            weave_parameters WP2860_2860
0.0000 0.0000 0.0000
      BISECTOR_FRAME bf_23990 (euler 92.4159 109.8739 95.2751)
      VECTOR (euler -62.3807 76.3037 -101.7527)
            parameter_list PL20_20
            weave_parameters WP20_20
;

to W37170_37170  add S37170_26890
    (shift X -2.1131 Y 133.8987 Z 157.4915
    euler 0.0000 0.0000 0.0000);

{ Seam - weld curve : entity #30560 }
SEAM S37170_30560
   PRISM_SIDES 3 PRISM_RADIUS 1.0 VECTOR_LENGTH 20.0
    list_template LT2890_2890  weave_template WT2890_2890
   COMPONENT POLYLINE
0.0000 0.0000 0.0000
      BISECTOR_FRAME bf_24710s (euler 90.6222 110.0124 92.0967)
      VECTOR (euler -63.6782 75.2433 -99.1247)
            parameter_list PL2910_2910
            weave_parameters WP2910_2910
...
0.0000 0.0000 0.0000
      BISECTOR_FRAME bf_24710 (euler 90.6222 110.0124 92.0967)
      VECTOR (euler -63.6782 75.2433 -99.1247)
            parameter_list PL2910_2910
            weave_parameters WP2910_2910
;

to W37170_37170  add S37170_30560
    (shift X 0.0000 Y 134.5000 Z 150.2704
    euler 0.0000 0.0000 0.0000);
```

```
{ Seam - weld curve : entity #34230 }
SEAM S37170_34230
    PRISM_SIDES 3 PRISM_RADIUS 1.0 VECTOR_LENGTH 20.0
    list_template LT2890_2890  weave_template WT2890_2890
    COMPONENT POLYLINE
0.0000 0.0000 0.0000
      BISECTOR_FRAME bf_25430s (euler 91.7380 109.8966 96.0215)
      VECTOR (euler -63.1751 76.5005 -102.3988)
            parameter_list PL5790_5790
            weave_parameters WP5790_5790
...
0.0000 0.0000 0.0000
      BISECTOR_FRAME bf_25430 (euler 91.7380 109.8966 96.0215)
      VECTOR (euler -63.1751 76.5005 -102.3988)
            parameter_list PL5790_5790
            weave_parameters WP5790_5790
;

to W37170_37170  add S37170_34230
    (shift X 0.0000 Y 141.2949 Z 154.9984
    euler 0.0000 0.0000 0.0000);

{ Weld Seam Data  }

set W37190_37190  = ;

seam_list
   default white white white white
   S37170_34230
   S37170_30560
   S37170_26890
;
```

Annex 6 Examples of IRL Exchange Files in InterRob

P. Sorenti
BYG Systems Ltd, William Lee Building, Highfields Science Pard
Nottingham NG7 2RQ, United Kingdom

Example IRL welding program generated by GRASP

The following IRL program was generated in GRASP to perform a simple welding task between two points with weaving. The program illustrates the syntax of welding channel data to control the welding process, a new feature added to the IRL programming syntax for arc welding technology.

```
PROGRAM WELDER2;

VAR
{ Global variables used for Input/Output }
  OUTPUT REAL : VOLTAGE AT 1, POWER AT 3 ;

BEGIN

MOVE LIN
  ROBTARGET(  POSE(
              POSITION(-1.3977395,310.1186218,-685.4429321),
                          ORIZYX(180.0,.0,180.0)),
              -4.07,312.06,-205.33,90.0,.0,.0,
              ADD_JOINT(.0,89.9999847))
              SPEED := 5.0 ;

WEAVING_PATTERN(1.5,1.7,35.0) ;
WEAVING_PLANE(-1.4,310.12,-785.44) ;
WEAVING_ON() ;

WELDING_ON() ;

VOLTAGE := 6.1 ;
POWER := 6.2 ;

{ -----> Position step <----- }

MOVE LIN
  ROBTARGET(  POSE(
              POSITION(-301.3977356,310.1186218,-685.4429932),
                          ORIZYX(180.0,.0,180.0)),
```

```
              -304.07,312.06,-205.33,90.0,.0,.0,
              ADD_JOINT(.0,89.9999847))
              SPEED := 5.0 ;

WEAVING_OFF() ;
WELDING_OFF() ;

ENDPROGRAM ;
```

References

Bernhardt, R. et al. (1994) The Realistic Robot Simulation Interface Specification. Version 1.0. January 19, 1994. (Ed. Dr. R. Bernhardt), IPK-Berlin, Pascalstrasse 8-9, 109 587, Berlin, Germany

Bey, I. et al. (Eds.) (1994) Neutral Interfaces in Design, Simulation, and Programming for Robotics. Research Reports ESPRIT. Subseries PDT. Project 5109 NIRO. Springer, Berlin

Bottema, O. and Roth, B. (1979) Theoretical Kinematics, North-Holland, Amsterdam

Cattell, R.G.G. (1991) Object Data Management – Object-Oriented and Extended Relational Database Systems, Addison-Wesley, New York

DIN (1990) DIN 66 313: IRDATA - Interface between programming system and robot control. Berlin: Beuth-Verlag

DIN (1994) DIN 66 312: IRL - Industrial Robot Language. Teil 1 and 2, Version 1.2, (First version published in Berlin: Beuth-Verlag 1992)

Goldberg, A. and Robson, D (1983) Smalltalk 80 – The Language and its Implementation, Addison-Wesley, New York

Horsch, T. et al. (1994) Prototype software. Information can be obtained from Dr. T. Horsch, Reis Robotics, P.O.Box 11 01 61, D-6377 Obernburg

Horsch, T. et al. (1995) Prototype software. Information can be obtained from Dr. T. Horsch, Reis Robotics, P.O.Box 11 01 61, D-6377 Obernburg

Hoschek and Lasser (1993) Fundamentals of Computer Aided Geometric Design, AK Peters, Wellesley

ISO (1991) ICR - Intermediate Code for Robots. Manipulating Industrial Robots, ISO DP 10562-2

ISO 10303-11 (1994) Industrial automation systems and integration – Product data representation and exchange – Part 11: Description methods: The EXPRESS language reference manual

ISO 10303-203 (1994) Industrial automation systems and integration – Product data representation and exchange – Part 203: Application protocol: Configuration controlled 3D designs of mechanical parts and assemblies

ISO 10303-21 (1994) Industrial automation systems and integration – Product data representation and exchange – Part 21: Implementation methods: Clear text encoding of the exchange structure

ISO 10303-41 (1994) Industrial automation systems and integration – Product data representation and exchange – Part 41: Integrated generic resources: Fundamentals of product description and support

ISO/CD 10303-214 (1995) Industrial automation systems and integration – Product data representation and exchange – Part 214: Application protocol: Core data for automotive mechanical design processes. Document number ISO TC184/SC4 N319 and ISO TC184/SC4/WG3/N436

ISO/DIS 10303-105 (1995) Industrial automation systems and integration – Product data representation and exchange – Part 105: Integrated application resources: Kinematics.

Jüttler, B., (1993) Über zwangsläufige rationale Bewegungsvorgänge, Sitzungsbericht der Österreichischen Akademie der Wissenschaften. 202, 117-132

Ludwig, A. (Ed.) (1993) ProSTEP Integrated Model Schema, Version 2.1. Unpublished report Kf9312.605, Kernforschungszentrum Karlsruhe GmbH

Nielsen, L.F. (1994) Dynamic Simulation of the Reis RV6 Robot. Control Engineering Institute, Technical University of Denmark, IfS Report no. 894.55. InterRob.WP2.DTU.06.94

ObjectStore Release 3.0 (1993) User Guide. Library Interface, Object Design, Inc.

Owen, J. (1993) STEP - An Introduction. Information Geometers, UK

Pedersen, T. (Ed.) (1995) Specification of database implementation. InterRob (ESPRIT Project 6457) Deliverable No: D3.2.1

Schenck, D. and Wilson, P. (1994) Information Modelling - The EXPRESS Way. Oxford University Press, Inc.

Schlechtendahl E.G. (Ed.) (1989) CAD Data Transfer for Solid Models. Research Reports ESPRIT. CAD Interfaces (CAD*I), Vol. 3, Berlin: Springer

Sørensen, T. (Ed) (1993) Identification of Geometry, Kinematics, Dynamics, Control, and Robotics Specific Data, ESPRIT Project 6457, InterRob. Deliverable D1.1.1. InterRob.WP1.DTH.10.93, IFS Rapport nr. S93.72, Control Engineering Institute, Technical University of Denmark

Sørensen, T. (Ed.) (1994) Pilot Specification of a STEP Based Reference Model for Exchange of Robotics Models. ESPRIT Project 6457, InterRob. Deliverable D1.1.2. InterRob.WP1.DTU.06.94, IFS Rapport nr. S94.40, Control Engineering Institute, Technical University of Denmark

Sørensen, T. (Ed.) (1995) Final Specification of a STEP Based Reference Model for Exchange of Robotics Models. ESPRIT Project 6457, InterRob. Deliverable D1.1.3. InterRob.WP1.DTU.06.95, IFS Rapport nr. S95.23, Control Engineering Institute, Technical University of Denmark

Sørensen, T. (Ed.) (1996) Final specification of a STEP based reference model for exchange of robotic models InterRob. FZKA-PFT 174, Forschungszentrum Karlsruhe/D

STEP Programmer's Toolkit (1992) Version 1.2, ROSE Library Reference Manual, STEP Tools, Inc.

STEP Programmer's Toolkit (1992) Version 1.2, STEP Interface to ObjectStore – User's Guide, STEP Tools, Inc.

Stroustrup, B., (1991) The C++ Programming Language, 2nd edition, Addison-Wesley, New York

Wapler, M. (1993) PRODEX - Product Model Exchange Using STEP. ESPRIT Project 6040. Fraunhofer-Gesellschaft IPA, Stuttgart

Research Reports ESPRIT

Area *Computer-Integrated Manufacturing and Engineering (CIME)*

Improving the Performance of Neutral File Data Transfers. Edited by R.J. Goult, P.A. Sherar. IX, 138 pages. 1990 (Project 322 CAD*I, CAD Interfaces, Vol. 6)

Advanced Modelling for CAD/CAM Systems. Edited by H. Grabowski, R. Anderl, M.J. Pratt. VI, 113 pages. 1991 (Project 322 CAD*I, Vol. 7)

IMPPACT Reference Model. Edited by W.F. Gielingh, A.K. Suhm. XII, 261 pages. 1993 (Project 2165 IMPPACT, Integrated Modelling of Products and Processes using Advanced Computer Technologies)

CIMOSA: Open System Architecture for CIM. Edited by ESPRIT Consortium AMICE. XI, 234 pages. 2nd, rev. and ext. edition 1993 (Project 688/5288 AMICE, A European CIM Architecture)

Vibration Control of Flexible Servo Mechanisms. Edited by J.-L. Faillot. VII, 206 pages, 1993 (Project 1561 SACODY, A High Performance Flexible Manufacturing System (FMS) Robot with On-Line Dynamic Compensation)

CCE: An Integration Platform for Distributed Manufacturing Applications. A Survey of Advanced Computing Technologies. Edited by ESPRIT Consortium CCE-CNMA. XII, 207 pages. 1995 (Project 7096 CCE-CNMA, CIME Computing Environment: Integrating CNMA, Vol. 1)

MMS: A Communication Language for Manufacturing. Edited by ESPRIT Consortium CCE-CNMA. XII, 185 pages. 1995 (Project 7096 CCE-CNMA, Vol. 2)

Subseries PDT (Product Data Technology)

CAD Geometry Data Exchange Using STEP. Edited by H.J. Helpenstein. XIV, 432 pages. 1993 (Project 2195 CADEX, CAD Geometry Data Exchange)

Neutral Interfaces in Design, Simulation, and Programming or Robotics. Edited by I. Bey et al. XV, 334 pages, 6 figs. 1994 (Project 2614/5105 NIRO, Neutral Interfaces for Robotics)

NEUTRABAS. A Neutral Product Definition Database for Large Multifunctional Systems. Edited by H. Nowacki. XII, 203 pages. 1995 (Project 2010 NEUTRABAS)

Computer Aided Concurrent Integral Design. Edited by R.F. Schmidt and M. Schmidt. X,78 pages. 1996 (Project 5168 CACID)

Interoperability of Standards for Robotics in CIME. Edited by F. Mikosch. IX,140 pages. 1997 (Project 6457 InterRob)

Springer
and the
environment

At Springer we firmly believe that an international science publisher has a special obligation to the environment, and our corporate policies consistently reflect this conviction.
We also expect our business partners – paper mills, printers, packaging manufacturers, etc. – to commit themselves to using materials and production processes that do not harm the environment. The paper in this book is made from low- or no-chlorine pulp and is acid free, in conformance with international standards for paper permanency.

 Springer